T0189125

Textile Science and Clothing Technology

Series editor

Subramanian Senthilkannan Muthu, Kowloon, Hong Kong

More information about this series at http://www.springer.com/series/13111

Subramanian Senthilkannan Muthu
Editor

Sustainable Innovations in Textile Chemistry and Dyes

 Springer

Editor
Subramanian Senthilkannan Muthu
Kowloon
Hong Kong

ISSN 2197-9863   ISSN 2197-9871   (electronic)
Textile Science and Clothing Technology
ISBN 978-981-13-4196-0   ISBN 978-981-10-8600-7   (eBook)
https://doi.org/10.1007/978-981-10-8600-7

Printed on acid-free paper

This Springer imprint is published by the registered company Springer Nature Singapore Pte Ltd. part of Springer Nature
The registered company address is: 152 Beach Road, #21-01/04 Gateway East, Singapore 189721, Singapore

This book is dedicated to:
The lotus feet of my beloved
Lord Pazhaniandavar
My beloved late Father
My beloved Mother
My beloved Wife Karpagam and
Daughters—Anu and Karthika
My beloved Brother
Last but not least
To everyone working in the global textile wet
processing sector to make it GREEN &
SUSTAINABLE

Contents

Chapter 1
Low Impact Reactive Dyeing Methods for Cotton for Sustainable Manufacturing

M. Gopalakrishnan, K. Shabaridharan and D. Saravanan

Abstract The reactive dyes, though result in higher fastness properties compared to other classes of dyes used in colouration of cotton fibres, lead to problems related to effluents predominantly due to unreacted and hydrolysed dyes during the reaction. Besides, use of higher amounts of salts added during the exhaustion (30–70 g/L) and fixation process (10–20 g/L) also result in higher COD levels in the effluents. Suitable modifications of dyes to improve the reactivity or to lower the hydrolysis and modification of substrates to have more reactive sites could provide a sustainable solution to such problems. Various scopes available for modifying the substrates suitable for reactive dye applications and different structural modifications carried out in the reactive dyes to reduce the environmental impacts are discussed in this chapter.

Keywords Chloro triazine · Covalent bond · Fastness
Hydrolysis · Substantivity · Surface modification

1 Introduction

The Global Fiber Consumption Survey 2016 conducted by the Lenzing reinstates the dominant position of cotton fibres, with the consumption of cotton fibre for apparels around 25% [1]. Majority of the apparels are still produced with cotton fibre for its softness and breathability. On the other hand, majority of cotton materials are dyed with reactive dyes for all-round fastness properties. Reactive dyes are preferred in the

M. Gopalakrishnan (✉) · K. Shabaridharan · D. Saravanan
Department of Textile Technology, Bannari Amman Institute of Technology,
Sathyamangalam, Erode District 638401, Tamil Nadu, India
e-mail: gokin_m@yahoo.co.in

K. Shabaridharan
e-mail: shabari.iit@gmail.com

D. Saravanan
e-mail: dhapathe2001@rediffmail.com

© Springer Nature Singapore Pte Ltd. 2018
S. S. Muthu (ed.), *Sustainable Innovations in Textile Chemistry and Dyes*, Textile Science and Clothing Technology, https://doi.org/10.1007/978-981-10-8600-7_1

case of cotton wherever the chlorine fastness is not demanded, significantly. Prior to the commercial applications of reactive dyes in the year 1956, cotton substances were dyed with direct dyes and, vat and acid dyes with certain pre-treatments. Substantivity of dyes eliminates the pretreatments and simplifies the dyeing process. Needless to state, the substantive reactive dyes satisfy the customer needs that the dyed substances would withstand severe washing and service conditions. Reactive dye is the one that forms a strong covalent bond with the side chain of cellulose that leads to good fastness properties. Reactive dyes are also recommended for their brilliancy and it's varieties in colour [2, 3].

Structure of the reactive dyes makes them more attractive, by covalently reacting with the substrates, which also leads to less impact onto the environment with reasonably better fastness properties. The characteristic features of a typical reactive dye molecule include, (i) chromophoric grouping, contributing the colour and much of the substantivity for cellulose; (ii) reactive system, enabling the dye to react with the hydroxyl groups in cellulose; (iii) bridging group that links the reactive system to the chromophore; and, (iv) one or more solubilising groups, usually sulphonic acid substituents attached to the chromophoric group. In general, reactive dyes react with hydroxyl groups present in cellulose and form covalent bond with cellulose under alkaline conditions that has been extensively dealt in the past. Mostly, one cellulosic molecule is attached to one reactive site in the reactive dye. However, sometimes two cellulosic molecules may attach to two reactive groups in dyes. Reactive dyes also react with hydroxyl groups present in water, leading to hydrolysis of the dyes and loss of dyes. These dead dyes cannot react with cellulose further and cause increase in the effluent loads. These hydrolyzed dyes not only leads to poor colour yield, this may also leads to poor fastness and needs severe washing and soaping treatments to remove the unfixed dyes. Reactive dyes are more substantive towards cellulose rather than water molecule, so 60–70% of the reactive dyes are exhausted on cotton fibres, in the original form, and 30–40% of dyes may react with water and leads to lower colour yield [4].

However, suitable modifications of the reactive dye structure further and surface of the reacting surface (fibres and fabrics) are often explored to improve the reactivity between fibres and dyes, thereby reducing the environmental impacts. Table 1 shows the various possibilities to improve the reactivity and reduce the amount of un-reacted dyes let into the effluents.

1.1 Types of Reactive Dye Systems

Table 2 gives the list of reactive dye systems, however, all these dyes are not used in practical applications and only a few reactive dye systems are used in the industry [5]. After the first invention of dichloro triazine reactive dyes by ICI, the less reactive mono chloro triazine dyes were developed to improve the dye bath exhaustion by reducing the hydrolysis. Cibacron F range of dyes is based on fluorine as leaving group and gives higher level of reactivity than 2 amino 4-chloro

Table 1 Sustainable routes to reduce the environmental impacts—reactive dyes applications

Process method	Reaction system	Possible impacts
Conventional process	Reactive dyes + Cotton fibres + Reaction medium	Hydrolysis of dyes in reaction medium Low reaction with substrates Unreacted dyes in effluent
Sustainable route	Modified reactive dyes + Cotton fibres + Reaction medium	More reactive sites in the dyes Better reaction (fixation) with substrates Low residual dyes in effluent Wider pH tolerant conditions Wider temperature tolerant conditions Less requirements of auxiliaries Less Wash-off cycles Possibility of dyeing blends in a single bath Higher fastness properties Less associated pollutants (free metals and others substances present in dyes and auxiliaries)
	Reactive dyes + Modified cotton fibres + Reaction medium	
	Modified reactive dyes + Modified cotton fibres + Reaction medium	

analogues. The reactivity of a Remazol, a vinyl sulphone reactive dye, having a precursor 2-sulphatoethylsulphonyl, is in the range between the high-reactivity heterocyclic systems, such as dichlorotriazine and the low-reactivity ranges, such as aminochlorotriazine or trichloropyrimidine [6, 7].

1.2 Homo Bifunctional (2MCT) Type (HE Brand) Reactive Dye

Reactive HE types of dyes, having two triazinyl dye groups, are categorized with low affinity (like M Brand Reactive dyes) and high exhaustion and fixation (unlike M Brand Reactive dyes). Since, the dye bath exhaustion is very high and the residual dye bath contains less hydrolysed dyes, washing-off of the dyeing goods are much easier. The monochloro-s-triazine dyes are less reactive than dichloro triazine dyes, so the stability of the dyes are high. Due to two monochloro triazine dyes present in the dye structure, if, one gets hydrolysed with water the chances of other to react with

Table 2 Important reactive dye system [5]

Monofunctional dyes	
Dye system	Typical brand name
Dichlorotriazine	Procion MX
Aminochlorotriazine	Procion H
Aminofluorotriazine	Cibacron F
Trichloropyrimidine	Drimarene X
Chlorodifluoropyrimidine	Drimarene K
Dichloroquinoxaline	Levafix E
Sulphatoethylsulphone	Remazol
Sulphatoethylsulphonamine	Remazol D
Bi-functional dyes	
Bis(aminochlorotriazine)	Procion H-E
Bis(aminonicotinotriazine)	Kayacelone React
Aminochlorotriazine–Sulphatoethylsulphone	Sumifix Supra
Aminofluorotriazine–Sulphatoethylsulphone	Cibacron C

cellulose under alkaline conditions are higher, which leads to higher shade build-up, leaving less hydrolysed dyes.

2 New Range of Reactive Dyes

One type of reactive dye cannot suit for all the applications methods due to different process conditions adopted in those processes. So, the dye manufacturers often develop a new range of reactive dyes for the various application conditions, for example, the dye which is used in exhaustive type of applications, low reactivity and high substantivity dyes are produced for the best results. Similarly, high reactivity ranges of dyes are developed for continuous (process) applications. The idea of single dye (structure) to meet the above needs, often necessitates higher quantities of dyes to compensate the losses due to hydrolysis, high addition of auxiliaries to increase the exhaustion and fixation of dyes. Such measures lead to higher pollution loads in the effluents and effluent treatment systems. Hence, it becomes imperative to modify either the structure of the dyestuff used in order to improve the exhaustion and withstand the process conditions or modification of the substrates involved in the application process to enhance the reactivity between the dye-fibre. In this section, some of the new range of high exhaustive reactive dyes, suitable for exhaust and continuous methods, that are commercially available in the market are listed.

2.1 New Range of Dyes for Exhaust Dyeing

Dye manufacturers continuously develop the new dyes to improve the sustainability, by reducing the environmental impacts. Today, new range of reactive dyes are available in the market with the aim of increasing the substantivity thereby reducing the dye wastage and reducing the effluent load. List of dyes and type, commercial name, manufacturers and the applications are listed out in the Table 3 [8, 9]. These dyes exhibits high fixation rates, much shorter washing-off cycles and hence reduced effluent loads. These dyes also display better exhaustion properties, even in low salt additions with good reproducibility capabilities.

2.2 New Range of Reactive Dyes for Continuous Dyeing

In the case of continuous application systems, chances of hydrolysis are expected to be very less and hence low substantivity and high-reactive, reactive dyes are preferred. Table 4 lists the new range of reactive dyes, which are available in the market, vary with respect to its chemical structure and the level of substantivity, ranging from medium to low [8, 9]. These dyes are expected to yield better performance in the continuous processes like padding method, pad patch method and printing operations, hold good in terms of reproducing capability of shades or colours, very short washing cycle and good reactivity, reduce the dyeing cost by reducing the washing cycles, increase the exhaustion and reducing the effluent loads and exhibit improved all-round fastness properties.

3 Modification of Reactive Dyes

Reactive dyes and cellulose are anionic nature, in aqueous medium, and both repel each other. Moreover, the commercial reactive dyes have poor exhaustion properties due to lack of substantivity and the chances of hydrolysis are more in such dyes. In this section, some of the methods to modify the commercial reactive dyes to overcome these deficiencies are highlighted.

3.1 Modification of Vinyl Sulfone Reactive Dye

Initial vinyl sulfone reactive dyes had β-sulfatoethylsulfone group, a non-reactive group with necleophiles. During dyeing, β- group eliminates and form free vinyl sulfone reactive dye, which is reactive with nucleophiles (water, amines, alcohol and cellulose) and forms covalent bond, by nucleophilic addition reaction, with the

Table 3 Improving dye fixation to enhance the reactivity [8, 9]

S. No.	Dye name (Commercial name)	Dye type	Improvements in the reaction system
1	Huntsman–AVITERA® SE	Poly-reactive dyes Reactive dyes (60 °C exhaust dyeing)	A high fixation rate Much shorter washing-off cycle Reduce water consumption and save energy
2	Huntsman–NOVACRON® LS	Innovative bi-reactive dyes Reactive dyes (70 °C exhaust dyeing)	Require only 25% salt that is used in conventional reactive dyes Have very strong build-up and high fixation, resulting in outstanding reproducibility and less pollution. Can be applied using one-bath dyeing on polyester/cellulose (PES/CEL) blends, saving dyeing time
3	Huntsman–NOVACRON® FN	Reactive dyes (60 °C exhaust dyeing)	Very high solubility, good diffusion and levelness, and high fixation. Suitable for short-LR dyeing, with outstanding compatibility, excellent reproducibility, easy washing-off and good all-round fastness
4	Huntsman–NOVACRON® S	Reactive dyes (high reactivity; 60 °C exhaust dyeing) cold pad batch dyeing of dark shades	Developed for medium-to-dark shades. Exhibits outstanding build-up and can achieve very dark shades. Brown, Bordeaux and Black formulations deliver substantial cost advantages Provides top reproducibility and easy washing-off
5	Huntsman–NOVACRON® TS	Reactive dyes (60 °C exhaust dyeing)	Optimizing cost while fulfilling market demand for fastness, Recommended for medium to dark shades Offer consistent levelness and repeatable results

(continued)

Table 3 (continued)

S. No.	Dye name (Commercial name)	Dye type	Improvements in the reaction system
6	Huntsman–NOVACRON® W	Reactive dyes (high reactivity; 60 °C exhaust dyeing)	Developed for medium-to-dark shades Exhibits outstanding build-up and can achieve very dark shades. Brown, Bordeaux and Black formulations deliver substantial cost advantages. Provides top reproducibility and easy washing-off
7	Jay chemicals–Jakofix HE	Homo bi-functional dyes	Economical products Suitable for post mercerizing and post-bleaching
8	Jay chemicals–Jakofix Supra HR	Homo bi-functional dyes	Optimum reproducibility under difficult dyeing conditions and in blends with polyester for RFT levels Excellent leveling properties under difficult dyeing conditions such as winch and cabinet dyeing machines and on difficult fabrics such as viscose/Lycra blends or garments with thick seams Resistant to repeated domestic washing Resistant to perborate wet fading Good perspiration and light fastness
9	Jay chemicals–Jakazol LD	Hetero bi-functional dyes	Easy clean-down process suitable for short runs Excellent compatibility and reproducibility for high Right First Time levels Good level dyeing properties Easy to wash-off

(continued)

Table 3 (continued)

S. No.	Dye name (Commercial name)	Dye type	Improvements in the reaction system
10	Jay chemicals–Jakazol CE	Hetero bi-functional dyes	Easy clean-down process suitable for short runs Reduced tailing Good build-up for medium-heavy shades Good reproducibility for high RFT levels Suitable for warm exhaust, Pad-dry-chemical pad steam, Pad-dry-steam, Pad-dry-thermofix and E-control
11	Jay chemicals–Jakazol DS	Poly functional dyes	Easy clean-down process suitable for short runs Very high strength products Excellent build-up for very deep shades Good wash-off and fastness levels Resistant to perborate wet fading Resistant to repeated domestic washing Process suitable for warm exhaust, pad-dry-chemical pad steam, pad-dry-steam, pad-dry-thermofix and E-control
12	Jay chemicals–Jakazol VS	Vinyl sulphone dyes	Easy clean-down process suitable for short runs Wide range of products for broad shade gamut Multiple options for economical black and navy shades Range of dischargeable dyes for ground shades Good wash-off properties for good fastness levels Process suitable for warm exhaust, pad-dry-chemical pad steam, pad-dry-steam, pad-dry-thermofix, E-control

(continued)

Table 3 (continued)

S. No.	Dye name (Commercial name)	Dye type	Improvements in the reaction system
13	Jay chemicals–Jakazol HLF	High light fast dyes	Easy clean-down process suitable for short runs Good wash-off properties for good fastness High light fastness in pale shades Reproducibility giving good RFT levels compatibility with level dyeing performance No available free metal Process suitable for warm exhaust, pad-dry-chemical pad steam, pad-dry-steam, pad-dry-thermofix, E-control

chances of hydrolysis in reaction water around 40%. Modified dyes with a cross-linking agent disulfine-bis-ethylsulfone (DSBES) increase the substantivity of the reactive dyes temporarily and during the alkaline fixation step, the dyes eliminate β- group and form two small vinyl sulfone reactive dyes and thus it improves the exhaustion and allow the unfixed dyes to wash off easily [10].

For preparation of disulfide-bis-ethylsulfone diamine intermediate, 1-Aminobenzene-4-S—thiosulfatoethylsulfone (Bunte salt) is dissolved in distilled water at pH 7 (using alkali), with thioglycollic acid at room tempearature for two hours. The precipitates are washed with distilled water and dried in an oven. Subsequently, the disulfide intermediate is dissolved in the concentrated sulfuric acid with the addition of sodium nitrite solution in a drop wise manner, while maintaining the temperature below 5 °C. Sulphamic acid is added to remove the excess nitrous acid and the solution is added into previously dissolved neutral solution of 1-naphthol 3, 6-disulfonic acid by water at below 4 °C. After 2 h, sodium chloride is added to salt-out the orange dye, which is then washed and dried at room temperature.

Dyeing of DSBES dye involves the regular reactive dyeing procedures with salt addition. The dyed fabrics exhibit very good primary exhaustion even at lower salt concentrations and need higher salt concentrations for secondary fixation stage.

3.2 Pyridinium Based Model Cationic Reactive Dye

The synthesis of the pyridinium based model cationic reactive dye involves addition of previously prepared solution of cyanuric chloride and acetone into the

Table 4 Reactive dyes recommended for high exhaustion in continuous dyeing [8, 9]

S.No.	Dye name	Dye type	Impact on processing
1	Huntsman–NOVACRON® C	Bi-reactive dyes (high reactivity, low-to-medium substantivity)	Provides high fixation rate, high reproducibility Fewer water effluent treatment problems and good fastness performance
2	Huntsman–NOVACRON® S	Reactive dyes (high reactivity; 60 °C exhaust dyeing)	Developed for medium to dark shades Exhibits outstanding build-up and can achieve very dark shades, including Brown, Bordeaux and Black Provides high reproducibility and easy washing-off
3	Huntsman–NOVACRON® TS	Reactive dyes (60 °C exhaust dyeing)	Recommended for medium to dark shades Offer consistent levelness and repeatable results and help maximize plant efficiency and productivity
4	Huntsman–NOVACRON® Brilliant Yellow EC-4G	High performance reactive dye	Offers a brilliant lemon yellow element, designed for dyeing at very short liquor ration Has a very good washing-off and very good overall fastness Has a good combinability with other NOVACRON® EC dyes
5	Huntsman–NOVACRON® P, NOVACRON® P Liq.	Reactive dyes (low affinity; printing)	Have low affinity and very good washing-off characteristics, resulting in reduced back staining
6	Jay chemicals–Jakazol CE	Hetero bi-functional dyes	Easy clean-down process suitable for short runs Reduced tailing and listing Good build-up for medium-heavy shades Good reproducibility for high RFT levels Process suitable for warm exhaust, pad-dry-chemical pad steam, pad-dry-steam, pad-dry-thermofix, E-control

(continued)

Table 4 (continued)

S.No.	Dye name	Dye type	Impact on processing
7	Jay chemicals–Jakazol DS	Poly functional dyes	Easy clean-down process suitable for short runs high strength products Excellent build-up for very deep shades Good wash-off and fastness levels Resistant to perborate wet fading Resistant to repeated domestic washing Process suitable for warm exhaust, pad-dry-chemical pad steam, pad-dry-steam, pad-dry-thermofix, E-control
8	Jay chemicals–Jakazol VS	Vinyl sulphone dyes	Easy clean-down process suitable for short runs Multiple options for economical black and navy shades Range of dischargeable dyes for ground shades Good wash-off properties for good fastness levels Process suitable for warm exhaust, Pad-dry-chemical pad steam, pad-dry-steam, pad-dry-thermofix, E-control
9	Jay chemicals–Jakazol HLF	High light fast dyes	Easy clean-down process suitable for short runs Good wash-off properties for good fastness High light fastness in pale shades Reproducibility for Right First Time levels Good level dyeing performance No available free metals Process suitable for warm exhaust, pad-dry-chemical pad steam, pad-dry-steam, pad-dry-thermofix, E-control

mixture of purified 1, 4-diamino-5- nitroanthraquinone and acetone. The mixture is continuously stirred at room temperature for four hours. Once the reaction completes, solution of pyridinium chloride and distilled water is added into the above solution and precipitate is purified and dried.

The modified cationic dye is applied on cotton by exhaustion method without addition of salts. The cotton fabric is immersed in previously dissolved modified cationic reactive dye and kept for 10 min at room temperature, then the temperature of the bath is increased to 85 °C at the rate of 2 °C/m. After 10 min, sodium carbonate is added and continued dyeing for another 45 min at 85 °C, followed by post dyeing washing-off. Cotton fabrics dyed with the modified reactive dyes, introducing the pyridinium solubilizing group, show good exhaustion and fixation than normal reactive dyes and eliminate the addition of salt with high level wash fastness [11].

3.3 Modified Reactive Dye Bis-Dichloro-S-Triazine Dye (Bis-DCT)

The intermediate, bis-ethylenediamine is dissolved in water at pH 13 (deprotonation of amines increases the solubility) and then hydrochloric acid is used to lower the pH to 10. The dye solution is added into an ice-cold solution of cyanuric chloride and stirred mechanically. Once the pH is stable, it is lowered to 7 with hydrochloric acid and 6.4 with disodium hydrogen phosphate and sodium dihydrogen phosphate. The mixture is filtered, centrifuged and washed to remove the residual cyanuric chloride and then dried. The resultant bis-dichloro- s-triazine dye is used to dye the cotton fabrics at 60 °C. The modified bis-dichloro-s-triazine dye exhibits good exhaustion than dichloro triazine dyes, around 90–95% [12, 13].

3.4 Modification of Dichloro Triazine Dyes

Initially, cotton fabrics are dyed with less substantive reactive dyes, mono chloro triazine, dichloro triazine dyes and these dyes exhibit poor exhaustion properties due to higher proportion of hydrolysed dyes. Some researchers reported the modified dyes with high exhaustion rates and one such method to modify the dichloro triazine dyes is to combine the two dichloro triazine dyes. The modified reactive dyes are prepared with combining the two commercial DCT dyes with 2 eq. of cysteamine, by 2 eq. of cyanuric chloride or 1 eq. of cysteamine followed by 1 eq. of cyanuric chloride [14].

Application of the modified dyes is carried out with regular dyeing procedure with salt and sodium carbonate. The modified dyes exhibit complete exhaustion than commercial reactive dyes at faster rates with ~25% higher exhaustion at 90 °C with salt than original commercial dyes [15, 16].

4   Chemically Modified Cotton–Cationic Cotton

Cotton substances are mostly dyed with reactive dyes due to their good fastness properties and brilliance in shade. But, dyeing of cotton with reactive dyes involves huge amount of electrolytes to increase the substantivity of the reactive dyes on cotton because, both reactive dyes and cotton (cellulose) are anionic in nature and repel each other in aqueous medium [17, 18]. So the reactive dyes have less affinity towards cotton in aqueous medium. Huge amounts of electrolyte (sodium chloride or sodium sulphate) are used to change the polarity of the cotton cellulose from anionic to cationic temporarily to overcome this lack of affinity of reactive dyes. Moreover, the amount of unfixed dye is more (around 30–40%) even with the addition of electrolytes (80–100 gpl), due to the possibility of hydrolysis of reactive dyes with water. If the dyes or cellulose is made cationic, both can attract each other and the effluent load will be very less (low salt). On the other hand, the amount of unfixed dye in cationised cotton is also less so the removal of colour becomes easier during the treatment of waste water [19]. In this section, the methods to produce the cationisation of cotton chemically before dyeing as a pretreatment to increase the affinity of the reactive dye towards cellulose, thereby, reducing the amount of electrolyte and colour in effluent are reported.

4.1   Oxidation of Cellulose

The charge of cotton is mainly depends on its chemical structure and the acidic group dissociation. Oxidation process changes the content of accessible acidic groups and content of acidic groups on the surface of cellulose. A mild oxidation process selectively introduces the carboxylic groups at C2 and C3 positions in the cellulose chain. Oxidation is carried out on cotton fabrics with 0.01 M aqueous solution of potassium periodate at 20 °C for 6 h, which introduces the aldehyde groups at C2 and C3 positions, followed by selective oxidation of the aldehyde groups by treating the sample further with 0.2 M sodium chlorite at 20 °C for 24 h. This process improves substantivity of the cotton and oxidation process has been suggested before cationisation, to improve the efficiency of cationisation process [20].

4.2   Cationic Agents for Cotton Pretreatments

4.2.1   Chitosan—The Polycationic Agent

The monosaccharide in cellulose polymer is D-glucose whereas, the chitosan has 2-amino-2-dehydroxy D-glucose [21]. The presence of amino group in the chitosan polymer makes the polymer soluble in diluted aqueous acidic solution as a poly cationic polymer while cotton is insoluble. This enables the cotton to be treated with polycationic polymer to improve the dyeability of cotton with anionic dyes like direct and reactive dyes. Chitosan based polycationic polymer is prepared by dissolving the chitosan with 1% aqueous acidic acid at 60 °C. The oxidized fabric samples are immersed in the chitosan solution at 60 °C for 2 h under constant stirring and the resultant fabrics are washed repeatedly to remove the unreacted chitosan and then dried [22, 23].

4.2.2 Tertiary Amine Cationic Polyacrylamide (TACPAM)

TENG Xiaoxu et al. used tertiary amine cationic polyacrylamide (TACPAM) as cationic agent for their treatment. Cationisation is carried out on bleached cotton fabrics is padded with 2% tertiary amine cationic polyacrylamide (TACPAM) at 20 °C for 2 min dip time with 80% wet pickup and treated at 100 °C for 5 min [24]. The cationic agent treated fabrics are dyed with reactive dyes either hot brand or ramazol class of dyes. The dyeing of treated fabrics, is carried out with regular dyeing procedures without salt addition and the concentration of cationic agent influence the amount of cationic sites on the fabric, higher the concentration of cationic agent, higher the cationic sites. However, if the concentration of the cationic agent increases the particular limit (beyond 2%), the fixation of the reactive dye is reduced.

4.2.3 Quaternized Polyglykol Ether of Fatty Amine (SINTEGAL V7CONC)

The cationic agent Sintegal V7 concentrate is applied on cotton fabric in exhaust method by varying the concentrations of the cationic agent solutions (0, 5, 1 and 2 g/L) at 50 °C without salt for 30 min [25]. The treatment is continued for another 30 min after the addition of 10 g/L Na_2CO_3. The resultant cationized fabric samples are neutralized in a dilute acetic acid solution, rinsed and dried.

4.2.4 3-Chloro-2-Hydroxypropyltrimethylammonium Chloride

The reactive material 3-Chloro-2-hydroxypropyltrimethylammonium chloride, available at 65% with alkali, is used for producing the cationised cotton. In a typical process, cotton fabrics are padded with 3-Chloro-2-hydroxypropyltrimethylammonium chloride (65%) and 50% sodium hydroxide. The cloth is wrapped to avoid migration and reaction of carbon dioxide with alkali to neutralize and, kept for 24 h at room temperature and subsequently rinsed in water for several times and neutralized. Dyeing of the padded fabric is recommended for treatment with sodium carbonate (5 gpl) and sodium hydroxide (0.7 gpl) prior to the addition of dyes. Subsequently, the temperature of the dye bath is raised to 60 °C and dyeing is continued for 30 min at pH 5–6. The colour yield of modified cationic cotton is more in this process, than 50% of the untreated cotton fabrics and moreover and the addition of huge amounts of electrolyte, sodium chloride is dispensed with, for reactive dye applications. The colour fastness to washing and light of the cationised cotton samples dyed with reactive dye is almost similar to that of untreated cotton samples [26–30].

4.2.5 Polyepichlorohydrin-Amine Polymers

Cationic agents are synthesized by mixing epichlorohydrin and dimethylamine with 1: 3 ratio at 95 °C for 18 h and the purified a cationic agent epichlorohydrin-dimethyl-amine (ECH-amine) is applied on cotton fabrics by either exhaust method or pad-dry method. Cationic agent (4 gpl) and sodium hydroxide (4 gpl) are used in exhaust method with 1:25 material to liquor ratio for 50 min at boiling temperature, while in pad-dry method, cationic agent (25 gpl) and sodium hydroxide (10 gpl) are used. T S Wu and KM Chen reported low colour yield at lower concentration of cationic agents, which increases with increasing the concentrations of agent. After reaching the optimum point (4 gpl), the increase in colour yield is very less [31].

4.2.6 Poly(Vinylamine Chloride)

The reactive polymeric quaternary ammonium compounds, amines or amides are used to modify the cotton as pretreatment before dyeing substantially to improve the dye update [24, 32]. However, the problem in this type of pretreatment is the proper selection of dyes and level dyeing. Poly (vinylamine chloride) (PVAmHCl) provides the wide range of properties in chelation to polymeric dyes and PVAmHCl has significant proportion of cationic sites ($+NH_3Cl-$) and, nucleophilic sites present in the primary amino groups of PVAmHCl, evident for salt-free dyeing. PVAmHCl is applied on cotton in padding mangle with 5 gpl concentration at pH 7 using dihydrogen phosphate and sodium hydroxide. The pretreated fabric samples are dried and cured for 10 min at 100 °C. The dyeing of the pretreated cotton sample is followed with regular reactive dyeing procedure without salt addition. The dye uptake of the pretreated cotton increases with increasing the concentration of pretreatment up to 10 gpl with excellent fastness properties [33].

4.2.7 Amino-Terminated Hyperbranched Polymer

Highly branched polymer dendrimers and hyperbranched polymers are widely preferred for modifying the cotton to achieve the salt free dyeing with level dyeing. The rich amino groups present in the amino-terminated hyperbranched polymer is the best agent for cationisation for cotton to improve the reactive dyeing behavior cotton without salt. The three dimensional structure and good solubility revolves the issue of unlevel dyeing after pretreatment [34].

Amino-terminated hyperbranched polymer is prepared with diethylene triamine and the solution of methyl acrylate and methanol as per the procedure described by Feng Zhang et al. Amino-terminated hyperbranched polymer is applied on cotton fabric in a padding mangle with citric acid as a binder, dried and cured at 160 °C for 3 min. The treated cotton fabric is dyed with regular dyeing procedures (using reactive dyes) without the addition of salt, which exhibits 30% increase in colour strength than untreated cotton fabrics. However, both treated and untreated cotton samples show similar wash fastness rating while treated cotton fabrics exhibit good levelling properties [34].

In general, the chemically modified cotton fabrics using cationic polymers show excellent dye uptake and good fastness properties. Addition of salt is eliminated during dyeing for pretreated fabrics with cationic agents and reduces the effluent loads [35–37]. Though dyestuff modifications and cationisation of cotton increase the exhaustion and eliminate the addition of salt, these approaches are largely at research level and yet to be commercialized widely. However, cold pad batch process eliminates the addition of huge amounts of salt as exhausting agent and requires less amount of water to dye the fabrics when compared with exhaust dyeing process. Other features of these process include (i) low cost than compared with conventional dyeing processes, (ii) elimination of electrolytes and other special, (iii) excellent wet fastness properties, (iv) reduced energy and water consumption (v) good level dyeing, (vi) chances of hydrolysis are less. Kusters offers a variety of dyeing machines, ranging from semi continuous to continuous processing. In the new range, Kusters has introduced the concept of swimming roll ('S'-roll) systems, equipped with deflection-controlled rolls to control the linear pressure for achieving the uniform pressure throughout the fabric width [38–42].

5   Use of Organic Salts

Dyeing of cotton fibres and fabrics with anionic dyes involves huge amounts of inorganic electrolytes, sodium chloride or sodium sulphate and, increase the problems in the effluent treatments. Alternate organic exhausting agent tetra sodium edate and trisodium citrate have been suggested in place of sodium chloride. Abo Farha S. A. et al. and Nahed S.E. have reported the use of the organic salt sodium edate for dyeing of cotton with reactive dyes without alkali, and Reda M. El-Shishtawy et al. have analysed the effect of sodium edate on dyeing of cotton/wool blended fabrics with hetero bi-functional reactive dyes. They have also reported that lesser quantities of organic salts result in the highest exhaustion percentage than compared to sodium chloride and, the increase in the concentration of sodium carbonate increase the fixation with optimum value at 15 gpl. Yu Guan et al. have reported the effect of biodegradable organic salt, sodium salt of polycarboxylic acid in reactive dyeing of cotton with dyeing of the cotton fabrics with organic salt (polycarboxylic acid sodium salt) around 60 gpl with reactive dyes, with or without alkali. The colour strength of the samples dyed with reactive dyes using polycarboxylic acid as exhausting agent, was similar to that of samples dyed with sodium chloride as the exhausting agent. Yu guan et al. have reported that the high alkalescence of sodium edate causes hydrolysis of dyes and very difficult to control it [43–45].

6 Foam Dyeing

In the past it was, often, assumed that the water would be available abundantly and cheap, which misled the processing industry to use it enormously in various processes including scouring, bleaching, dyeing and finishing. But after realization of the worse effects of the effluents on environment, and emergence of advances in science the chemistry attempts have been made to work on the reduction of loads in the effluents and to find an alternate to the reduce water in processing industry. Foam dyeing is one such process involves very little amount of water to dye the textile materials. Moreover, the energy required to dry the material is also very less with literarily no discharge of dyes in foam dyeing. Hong Yu et al. (2014) analyzed the influence of foam properties on dyeing of cotton with reactive dyes. The surfactant and stabilizers are mixed with aqueous solution at 100 rpm for 3 min. Dye solution is toughly mixed with foaming agent and stabilizer. Dyeing of cotton is carried out in a padding mangle or by a printing method. The process for foam dyeing is Foam generation—application of foam—steaming at 148 °C for 4 min—soaping to remove the unfixed dyes then washing. Hong Yu et al. (2014) reported that foam dyeing increases the colour strength and uniform dyeing is obtained with 0.1–0.2 mm bubble size [46–48].

7 Conclusion

Though dyeing of fibres and fabrics is an essential process in the value addition and garment manufacturing, it also adds to higher loads in the environmental pollution and effluents. Regardless to innovative methods available in the treatment of effluents to reduce the impact on the environment many attempts are made to alter the substrates and dyestuffs themselves to improve the substantivity of the dyes and thereby reducing the effluent loads. Bifunctional and multi-functional reactive dyes are some of the innovative solutions developed by the dyestuff manufacturers in this direction, while novel efforts are made to modify the substrates to accommodate more dyestuff and let negligible amounts into the effluents with less auxiliaries used in the process. Conscious efforts by the processors to use the combinations of these innovative solutions at various stages of manufacturing can provide sustainable solution to the existing manufacturing practice and significant reduction in the environmental impact.

References

1. http://www.lenzing.com/en/investors/equity-story/global-fiber-market.html
2. Rattee ID (1984) Reactive dyes for cellulose 1953–1983. Rev Prog Color 14:50–57
3. Rattee ID (1969) Reactive dyes in the coloration of cellulosic materials. J Soc Dye Colour 85:23–31. https://doi.org/10.1111/j.1478-4408.1969.tb02849.x
4. Carliell CM, Barclay SJ, Shaw C et al (1998) The effect of salts used in textile dyeing on microbial decolourisation of a reactive azo dye. Environ Technol (United Kingdom) 19:1133–1137. https://doi.org/10.1080/09593331908616772
5. Shore J (1995) Cellulosics dyeing. Society of Dyers and Colourists
6. Chinta SK, Vijaykumar S (2013) Technical facts and figures of reactive dyes used in textiles. Int J Eng Manag Sci 4:308–312
7. Zuwang W (1998) Recent developments of reactive dyes and reactive dyeing of silk. Rev Prog Color Relat Top 28:32–38. https://doi.org/10.1111/j.1478-4408.1998.tb00117.x
8. http://www.jaychemical.com/reactive-dyes.php
9. http://www.huntsman.com/textile_effects/a/Products/Dyes/Cellulosics/Reactive%20dyes%20for%20Exhaust%20Processes
10. Lewis DM, Renfrew AH, Siddique AA (2000) The synthesis and application of a new reactive dye based on disulfide-bis-ethylsulfone. Dye Pigment 47:151–167
11. Srikulkit K, Santifuengkul P (2000) Salt-free dyeing of cotton cellulose with a model cationic reactive dye. Color Technol 116:398–402. https://doi.org/10.1111/j.1478-4408.2000.tb00017.x
12. Morris KF, Lewis DM, Broadbent PJ (2008) Design and application of a multifunctional reactive dye capable of high fixation efficiency on cellulose. Color Technol 124:186–194. https://doi.org/10.1111/j.1478-4408.2008.00140.x
13. Sugimoto Tadaaki (1992) Neutral-Fixing reactive dyes for cotton Part 2-Commercial reactive dyestuff and their classification. JSDC 108:497–500
14. Smith B, Berger R, Freeman HS (2006) High affinity, high efficiency fibre-reactive dyes. Color Technol 122:187–193. https://doi.org/10.1111/j.1478-4408.2006.00032.x
15. Zheng C, Yuan A, Wang H, Sun J (2012) Dyeing properties of novel electrolyte-free reactive dyes on cotton fibre. Color Technol 128:204–207. https://doi.org/10.1111/j.1478-4408.2012.00364.x
16. Fujioka SS, Abeta S (1982) Development of novel mixed reactive reactive dyes with system. Dye Pigment 3:281–294
17. Ameri Dehabadi V, Buschmann HJ, Gutmann JS (2013) Pretreatment of cotton fabrics with polyamino carboxylic acids for salt-free dyeing of cotton with reactive dyes. Color Technol 129:155–158. https://doi.org/10.1111/cote.12010
18. Teng XX, Shi JW, Zhang SF (2013) Impact of reactive dye structures on dyeing properties in salt-free reactive dyeing. Adv Mater Res 781–784:2716–2721. https://doi.org/10.4028/www.scientific.net/AMR.781-784.2716
19. Blackburn RS, Burkinshaw SM (2002) A greener approach to cotton dyeings with excellent wash fastness. Green Chem 4:47–52. https://doi.org/10.1039/B111026H
20. Fras L, Johansson LS, Stenius P et al (2005) Analysis of the oxidation of cellulose fibres by titration and XPS. Colloids Surf A Physicochem Eng Asp 260:101–108. https://doi.org/10.1016/j.colsurfa.2005.01.035
21. Singha K, Maity S, Singha M (2013) The salt-free dyeing on cotton: an approach to effluent free mechanism; can chitosan be a potential option? Int J Text Sci 1:69–77. https://doi.org/10.5923/j.textile.20120106.03
22. Bhuiyan MAR, Shaid A, Khan MA (2014) Cationization of cotton fiber by chitosan and its dyeing with reactive dye without salt. Chem Mater Eng 2:96–100. https://doi.org/10.13189/cme.2014.020402
23. Kitkulnumchai Y, Ajavakom A, Sukwattanasinitt M (2008) Treatment of oxidized cellulose fabric with chitosan and its surface activity towards anionic reactive dyes. Cellulose 15:599–608. https://doi.org/10.1007/s10570-008-9214-8

24. Teng X, Ma W, Zhang S (2010) Application of tertiary amine cationic polyacrylamide with high cationic degree in salt-free dyeing of reactive dyes. Chin J Chem Eng 18:1023–1028. https://doi.org/10.1016/S1004-9541(09)60163-4

25. Ristić N, Ristić I (2012) Cationic modification of cotton fabrics and reactive dyeing characteristics. J Eng Fiber Fabr 7:113–121

26. Arivithamani N, Dev VRG (2017) Cationization of cotton for industrial scale salt-free reactive dyeing of garments. Clean Technol Environ Policy 19:2317–2326. https://doi.org/10.1007/s10098-017-1425-y

27. Wang L, Ma W, Zhang S et al (2009) Preparation of cationic cotton with two-bath pad-bake process and its application in salt-free dyeing. Carbohydr Polym 78:602–608. https://doi.org/10.1016/j.carbpol.2009.05.022

28. Ren JL, Sun RC, Liu CF et al (2006) Two-step preparation and thermal characterization of cationic 2-hydroxypropyltrimethylammonium chloride hemicellulose polymers from sugarcane bagasse. Polym Degrad Stab 91:2579–2587. https://doi.org/10.1016/j.polymdegradstab.2006.05.008

29. Heinze T, Haack V, Rensing S (2004) Starch derivatives of high degree of functionalization. 7. Preparation of cationic 2-hydroxypropyltrimethylammonium chloride starches. Starch/Staerke 56:288–296. https://doi.org/10.1002/star.200300243

30. Subramanian Senthil Kannan M, Gopalakrishnan M, Kumaravel S et al (2006) Influence of cationization of cotton on reactive dyeing. J Text Apparel Technol Manag 5:1–16

31. Evans GE, Shore J, Stead CV (1984) Dyeing behaviour of cotton after pretreatment with reactive quaternary compounds. J Soc Dye Colour 100:304–315. https://doi.org/10.1111/j.1478-4408.1984.tb00946.x

32. Lewis DM, McIlroy KA (1997) The chemical modification of cellulosic fibres t o enhance dyeability. Rev Prog Color 27:5–17. https://doi.org/10.1111/j.1478-4408.1997.tb03770.x

33. Ma W, Zhang S, Tang B, Yang J (2005) Pretreatment of cotton with poly (vinylamine chloride) for salt-free dyeing with reactive dyes. Color Techonol 121:193–197. https://doi.org/10.1111/j.1478-4408.2005.tb00272.x

34. Zhang F, Chen Y, Lin H, Lu Y (2007) Synthesis of an amino-terminated hyperbranched polymer and its application in reactive dyeing on cotton as a salt-free dyeing auxiliary. Color Technol 123:351–357. https://doi.org/10.1111/j.1478-4408.2007.00108.x

35. Cai Y, Pailthorpe MT, David SK (1999) A new method for improving the dyeability of cotton with reactive dyes. Text Res J 69:440–446. https://doi.org/10.1177/004051759906900608

36. Wu TS, Chen KM (1993) New cationic agents for improving the dyeability of cellulose fibres. Part 2-pretreating cotton with polyepichlorohydrin-amine polymers for improving dyeability with reactive dyes. J Soc Dye Colour 109:153–158. https://doi.org/10.1111/j.1478-4408.1993.tb01547.x

37. Hauser PJ, Tabba AH (2001) Improving the environmental and economic aspects of cotton dyeing using a cationised cotton+. Color Technol 117:282–288. https://doi.org/10.1111/j.1478-4408.2001.tb00076.x

38. www.benningergroup.com

39. Khatri Z, Ahmed F, Jhatial AK et al (2014) Cold pad-batch dyeing of cellulose nanofibers with reactive dyes. Cellulose 21:3089–3095. https://doi.org/10.1007/s10570-014-0320-5

40. Cpb T, Ag B The latest developments in the CPB dyeing process (Cold Pad Batch Process) Groz-Beckert : Five S of SEWING 5 stand, pp 40–41

41. Nitayaphat W, Morakotjinda P (2017) Cold pad-batch dyeing method for cotton fabric dyeing with Uncaria gambir bark using ultrasonic energy. Chiang Mai J Sci 44:1562–1569. https://doi.org/10.1016/j.ultsonch.2011.04.001

42. Khatri A, Peerzada MH, Mohsin M, White M (2015) A review on developments in dyeing cotton fabrics with reactive dyes for reducing effluent pollution. J Clean Prod 87:50–57. https://doi.org/10.1016/j.jclepro.2014.09.017

43. El-Shishtawy RM, Youssef YA, Ahmed NSE, Mousa AA (2007) The use of sodium edate in dyeing: II. Union dyeing of cotton/wool blend with hetero bi-functional reactive dyes. Dye Pigment 72:57–65. https://doi.org/10.1016/j.dyepig.2005.07.017

44. Farha SAA, Gamal AM, Sallam HB et al (2010) Sodium edate and sodium citrate as an exhausting and fixing agents for dyeing cotton fabric with reactive dyes and reuse of dyeing effluent. J Am Sci 6:109–127
45. Guan Yu, Zheng Qing-kang, Mao Ya-hong et al (2007) Application of polycarboxylic acid sodium salt in the dyeing of cotton fabric with reactive dyes. J Appl Polym Sci 105:726–732
46. Lewis DM (1993) New possibilities to improve cellulosic fibre dyeing processes with fibre-reactive systems. J Soc Dye Colour 109:357–364. https://doi.org/10.1111/j.1478-4408.1993.tb01514.x
47. Philips D (1996) Environmentally friendly, productive and reliable: priorities for cotton dyes and dyeing processes. J Soc Dye Colour 112:183–186. https://doi.org/10.1111/j.1478-4408.1996.tb01814.x
48. Zhang J, Zhang X, Fang K et al (2017) Effect of the water content of padded cotton fabrics on reactive dye fixation in the pad-steam process. Color Technol 133:57–64. https://doi.org/10.1111/cote.12253

Chapter 2
A Review of Some Sustainable Methods in Wool Dyeing

N. Gokarneshan

Abstract During the recent years, a good deal of research was carried out in developing natural dyes and mordants in dyeing of different natural fibres, which has paved the way towards achieving sustainability in the area of wet processing. This chapter highlights some significant trends in eco-friendly methods of wool dyeing. The influences of dye kind of mordant and dye concentration on the colour characteristics of dyed wool fibres have been studied. In order to quantify assessment of the influences of kind of mordant and concentration of dye, the calorimetric properties including colour strength, colour difference and colour coordinates have been considered. The findings reveal that the wool has great attraction for pinecone dye liquid, and mordant methods adopted show various shades between beige to brown which have good fastness. The reaction between diphenolic catechol and enzymes of potato juice has been used for optimization of wool colouration. No surplus chemicals are required for colour formation at low temperature. Investigation has been carried out on the effect of process variables such as temperature and corresponding concentration of catechol and plant juice on the colour intensity of the fabric. The temperature is found to significantly affect the colour strength. Wool dyed with different colours possess wash and light fastness. Fermented kum leaves have been applied on wool and comparative studies using traditional method of cold and heating process and chemical method using metallic salts have been investigated with regard to colour fastness properties, tensile strength and elongation percentage. The colour fastness in chemical as well as traditional methods exhibits fair to excellent results. The traditional hot method gives better results compared to the cold one. In order to study the colour fastness properties of colourant on wool fibres dyed with a natural dye extracted from the leaves of *Symplocus Racemosa* different mordants in combination with lemon juice have been used in suitable proportions. The test results of the dyed fabrics with regard to wash, rub, light and perspiration fastness have yielded fair to excellent fastness grades. Investigations have been carried out on kinetic and thermodynamic aspects of wool fabric using crude dye extract of *A. Nobilis* and

N. Gokarneshan (✉)
Department of Textile Technology, Park College of Engineering and Technology, Coimbatore, Tamil Nadu, India
e-mail: advaitcbe@rediffmail.com

compared with other natural dyes. Various dyeing parameters have been evaluated. All the dyeing methods discussed herein are good attempts to achieve sustainability. Each dyeing method offers its own merits besides achieving good properties. The advantages include optimization of dyeing process, better dyeability, affinity, flexibility in dyeing with other fibres besides wool, cost economy, improvement in properties, better utilization of natural resources, avoidance of harmful chemicals, etc.

Keywords Wool dyeing · Dye uptake · Diffusion coefficient · Colorant · Mordant

1 Introduction

Oxidative dyes from colorless dye percussors have been utilised in the permanent coloration of human hair. Such dyes comprise of a dye precursor as well as an oxidative agent (2 component system). The oxidant initiates the combination of the dye precursor and formation of color compounds that are of great molecular size colored compounds which get anchored to the structure of the fibre [1]. Increasing awareness on human health has prompted natural materials as acceptable options to synthetic products in different end uses, particularly dyeing and chemical industries. As natural dyes are biodegradable and offer a broad range of light shades, they are found to be compatible with synthetic dyes [2–4]. Natural colourants which are usually called pigments or dye molecules are of plant, animal, or mineral origins. The natural dyeing technique of textile materials has its their roots in India. Manipur in india is well known for handlooms and handicrafts. Mature leaves and young shoots of Strobilanthescusia (Nees) *Kuntzi*(Kum) have been used to produce a unique blue-black and indigo coloredKum-dye. Kum is the most important plant used in the dyeing of clothes by various communities in Manipur [5–9]. An unique feature of natural dyes lies in the ability to produce a wide range of rich colors which tend to complement one another [10]. In India natural dyes are used for dyeing with traditional wool and woolen products. A number of drawbacks are associated with textile dyeing using natural dyes. In the areas of pharmaceuticals, cosmetics and food, Arnebianobilis has been traditionally used as an important source of red colour [11, 12]. The outer layer of human skin as well as a number of animals comprises of a protein called keratin, which is tough and insoluble. Keratins belong to a group of structural proteins and largely found in wool, hair, feather, hooves and fingernails. Despite the huge quantities of keratin wastes, keratins do also find some way in the global market [13]. For example, feather meals are used for animals [14]. Keratin based cosmetics are used for treatment of human hair and skin [15, 16]. Keratin materials have found their way into various areas of applications such as concrete, ceramics, fertilizers, fire-fighting compositions, wound healing, leather tanning, production of bio hydrogen and shrink proofing of wool [17–24]. The chapter highlights the recent trends in wool dyeing which have been focused towards sustainability in a number of ways. Attempt has been made to produce necessary shade in wool by means of optimization of process condition for the oxidation reaction. The dyeing properties of wool have been investigated by measurement of CIE values, color strength K/S, and wash and light fastness values. Efforts have been taken to revive certain natural dyes that proved versatile in dyeing with cotton, wool and silk. The draw backs related to the use of natural dyes in textile dyeing have been overcome by combining them with newer types of mordants. Certain natural dyes from herbal extracts suitable for wool also suited polyester and nylon as well, thereby widening the area of application on textile materials. In order to enhance the dyeability using acid and reactive dyes a newer method of kertain protein extraction could be cross linked to wool fabric.

2   Use of Plant Juice in Wool Dyeing

Coloration in textiles has been produced by use of oxidative laccase enzymes with small colourless aromatic compounds like diamines, amino phenols, aminonaphthols and phenols, which undergo further non enzymatic reactions [25–31]. Enzymes are richly found in vegetable juices. But there has been no report regarding the study of the process of colouration of fibres by enzyme rich plant juices till date. Great quantities of potato juice produced by starch industry go as waste [32]. The potato juice has abundant polyphenol oxidase enzyme (PPO) [33]. PPO also known as tyrosinase, is a bifunctional, copper containing oxidase that contains catecholase as well as creaolase activity, which causes the browning reaction in fruits and plants [34]. The phenomenon of browning has been well researched by biochemists and is caused by the oxidation and dehydrogenation of colourless polyphenols by PPO.

In nature, the initial reaction catalyzed by PPO yields reddish brown products based on orthoquinones [35]. Such reactive species being very reactive further undergo a series of non enzymatic reactions that result in insoluble black-brown melanin pigments [36–38]. The optimum pH and temperature of the enzyme activity is found to be 6.6 temperature is 40 °C respectively.

As melanins are considered the most stable and resistant among known biochemical durable and deep coloration of textiles is made possible by utilizing these natural reactions. Based on such knowledge, fresh juice is extracted from the potato and combined with catechol to develop a series of brown shades on wool fabric. The optimization of process conditions for the oxidation reaction has been studied in order to produce the necessary shade on wool fabric.

2.1   Discussion of the Findings

A range of in situ colors ranging between light pinkish brown to deep reddish brown has been synthesized on wool by treating it with a mixture of catechol and fresh potato juice under various process conditions [39]. The process imitates the browning phenomenon that occurs in nature. The chemical compounds in potato juice are capable of catalyzing the oxidation of phenolic compounds to create delocalized free radical as the oxidized intermediate. They oxidize phenolic compounds and form coloured quinones with sulphhydryl and amino groups of wool [17].

2.2 Spectroscopic Investigations

Spectroscopic studies reveal that catechol has a sharp peak at 284.5 nm while potato juice shows a broad peak, with λ_{max} value coinciding with that of catechol at 286 nm. It is due to the fact that potato juice also contains phenolic compounds similar to catechol that cause browning of the vegetable. The coloured complex also reveals the main peak at 286 nm that contributes to the complex conjugated structures in the melanin molecule [18]. Between 280 and 600 nm, the second broad peak has been noticed. A diffused pattern without sharp peak is also indicated by the reflectance spectra of the colored fabric. Absence of a sharp peak in the visible region is a characteristic of several coloured compounds found in nature. In this regard, the colour arises from the synthesis of compounds relating to the group of melanins [39]. Melanins are heterogenous copolymers with large complex conjugated aromatic structures comprising of phenylene and oxyphenylene units arising from the C-C and C-O coupling of phenols [19]. Since standard analytical techniques such as UV and visible light analysis do not provide any significant data relating to melanins, the proper structure of the compound is not known so far [21].

2.3 XRD Studies

Investigations have been carried out relating to the Xray diffraction spectra of dyed and undyed wool [39]. In the case of both the samples the 2θ peak identified at 20.5, revealing that the colorant molecules are randomly distributed and are irregular in their pattern. Owing to insitu synthesis of color, the crystallinity of wool fibre is not noticed.

2.4 Light and Wash Fastness

The fastness of melanin is supposed to be high since it is stable and insoluble. The findings from fastness tests show that wash fastness of such colours range between 4–5 (Excellent) and light fastness grade is 4 considering a scale of 1–8 [39].

2.5   Experimental Design

A number of process factors like concentration of components, duration and temperature of treatment affect the color development in textiles. Optimization is required to find the best possible combination of process variables with respect to conservation of energy as well as materials used. A statistical design of experiments is employed to determine the process parameters that influence the color synthesized by coupling of potato juice with catechol juice in wool. It is possible to achieve many colors by change of process variables [39]. Although the K/S values for test samples varies widely from 1.94 to 10.65, there is not much change observed in the a* and b*values among the samples, indicating a tone on tone increase occurs in deeper colored samples without any change in hue or shade. Maximum depth of color (K/S = 10.65) for a sample where the concentration of potato juice as well as the temperature of treatment is at the highest level. The next highest K/S value of 9.04 is achieved in sample where the highest concentration of catechol is used at the highest temperature of treatment. Lower values of K/S (1.94–2.67) are obtained when treatment was carried out at the least temperature, irrespective of the concentration of catechol and potato juice used. This indicates that all parameters individually as well as collectively play a role in determining the final color obtained on wool. Statistical analysis is carried out to study these interactions further.

2.6   Statistical Analysis

Results obtained from ANOVA show that the model is highly significant for the treatment.

The temperature of treatment shows the highest influence on color. As the temperature increases the color strength increases. With increase in temperature between 30 to 90 °C, increase in the concentration of potato juice between 15 to 25% concentration increases the color strength values [39].

2.7   Contour Plots

The prediction of a recipe required to achieve a specific shade of wool is obtained by contour plots obtained from the design expert software. The Fig. 1 depicts the plots. As the concentration of the potato juice increases, the color strength value increases as can be visualised seen from option 1 wherein the highest temperature level (90 °C), and the concentration of catechol is maintained the lowest. But, only with greater concentration (3.5%) of catechol the highest value of 10 of color strength can be achieved [39]. It is interesting to note from option 2 that whether the concentration of potato juice is 25% or 15%, the K/S value remains same at 30 °C.

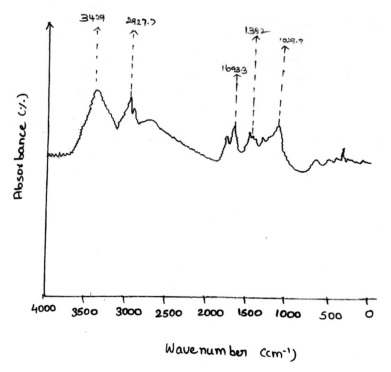

Fig. 1 FTIR relating to pine cone powder [53]

The third option shows that the complete range of colors from light to deep (K/S 2–10) can be achieved when the maximum concentration of potato juice (25%) is combined with minimum concentration of catechol-1%. As the temperature of treatment increases, the depth of color increases [39]. Even the maximum concentration of the two reactants, when used at the lowest temperature (30 °C), yields only a very light shade. It shows that deep shades can only be achieved at greater temperatures, regardless of the concentration of the reactants used. It is possible to achieve a broad range of shades by varying the temperature from 30–90 °C with various concentrations of catechol.

The synthesis of colorant does not occur at the temperature of 30 °C and whatever color is seen can be because of the inherent color of catechol itself. Hence, 30 °C should be avoided and only more than 45 °C should be used treatment temperature. It is possible to choose the most optimum process parameters for achieving a specific depth of color on wool.

3   Use of Natural Byproduct Colorant

Natural materials are beginning to supersede their synthetic counterparts due to increasing health awareness relating to many end uses including dyeing and chemical industries. Owing to the biodegradable nature and broad range of light shades achievable, natural dyes are compatible with their synthetic counterpart [40–42]. Natural colorants are derived from plants and animals. They are also known as pigments or dye molecules. Since natural dyes possess many considerable properties and also have versatility in application on various textile materials, they find use in textile industries.

Recently, many studies were developed for identifying these special natural resources which are very useful textile and related sectors involving coloring. One of the important economical natural sources are agricultural and forest wastes [42–44]. Based on the color index, flavonoid compounds are considered to be very popular among natural dyes applied on textile materials [45]. Pinecone is one such natural source abundantly available [46]. There are phenolic benzoic units in the chromophore of pinecone colorant and are partially linked to one another by means of pi-electrons [47]. The use of pinecones in separation of cationic and acid dyes has been studied by many researchers. However, the use of the pine cone as a natural dye in wool dyeing has been studied to a very limited extent [48–50].

The black pine (*Pinusnigra*) which is a pine species, is found in major parts of Iran [51]. Pinecones exist in forests as well as parks and are extracted from the big ornamental green tree. It is a natural colorant having a rounded spiral structure of length between 5–10 cm. Two types of cells of pinecone comprise of a number of chemical compounds in their wall cells that have a number of polar organic functional groups [52]. Usually, the pinecones are found in forests and are seen as disposable matter, that are gathered and despatched for disposal.

Studies have been carried out on the color characteristics of pinecone dyed wool. It is considered as an agricultural waste in Iran. Hence, pinecone powder has been used in various dyeing processes with three different concentrations based on fibre weight. In order to obtain dyed samples of acceptable fastness, deep shade wool has been premordanted good color fastness, high shade and variety of color shades.

During the process of dyeing salts of commercial metals are utilized. In order to investigate the dyeing properties a number of parameters have been determined.

Studies on pinecone show that it can be utilized in wool dyeing. When wool is dyed with pinecone a wide range of fast colors can be achieved. It has been found that a mordant concentration of 5% on fibre weight yields wool shade that is good. The dye uptake shows an increase at much concentrations of dye on fibre weight. It has been found that there is an enhancement in color and fastness properties in the premordanted wool having various mordants. By use of ferrous and copper sulphate the best dye uptake and fastness properties can be achieved. Use of pinecone powder in wool dyeing development is worth exploring due to its abundant properties.

3.1 Chemical Structure

FTIR spectrum of pinecone has been analyzed. The findings reveal that the indicative wavelength of hydroxyl group without bonding is at 3100–3600/cm and the carbohydroxyl group stretching is at 2927/cm. At a wavelength of 1639/cm the stretching of the carbonyl group can be seen and at wave length of 1447/cm the carboxylate anion observed [53]. At a wavelength of 1382/cm, a peak is observed due to the bending of the methylene group(C-H), and at a wave length of 1265/cm arises the phenol peak due to the stretching of the C-O group. An absorption peak of 1030/cm is observed in the linkage of the C-O-C ester group. Thus, FTIR spectra study reveals that the pinecone colorant extract consists of the presence of the chemical groups mentioned herein.

3.2 Measurements of Color

3.2.1 CIE Lab

The mordants have to be fixed on the surfaces of fibre for dyeing with natural dyes. The brightening mordants considered are stannous chloride, alum stannous chloride and potassium dichromate and the dulling mordants considered are cupric and ferrous sulfate. The lightness value, chroma and hue determine the distinct appearance of each colour [53]. Colour outcomes of dyed as well as undyed samples were measured and evaluated.

Crude wool exhibits a light yellow shade depending on the value of the angle of hue. The color tone is transformed from yellow to pale brown since these are dyed with powdered pinecone. The influence of greenness is lessened whereas that of redness increases as suggested by the color change behaviour.

The use of various mordant types can enable to obtain many colors from pinecone colorant as pointed by study. Evidently, the final colour, fastness characteristics and brightness are not only determined by the dye alone but also the concentration of the used mordants. The dye absorption is not satisfactory in the case of wool mordanted with 3% on fibre weight of metal salts as proved by the study. Moreover, different mordants exhibit similar shade.

Optimum absorbency of dye is achieved on the surface of wool dyed without mordants as clearly indicated by the lightness of colour. When the dye bath consists of 50% on fibre weight of pine colorant, the dye absorbability value is found to be high for without use of mordant free dye, since the attraction for pinecone dye molecules is indicated by lightness value. Test results show that in the case of treatment using different mordants the value of lightness reduces for the complete test sample. The least value of lightness of color are exhibited by the ferrous and copper sulfate in the dyed sample. Stannous chloride yields greater brightness of colour compared with any other mordant, while use of ferrous sulfate as mordant could lead to darkening and browning of the color shade. Lighter shades for wool are obtained from the mordants having greater values of L^*, whereas darker shades are achieved with mordants with a lower value of L^*. Using iron as the mordanting agent turns wool in black colour as pointed by test results. With increase in dye concentration the chromaticity of premordanted sample reduces. The fibre chromaticity can decrease by the use of some mordants such as copper and ferrous at 50 and 100% fibre weight of dye.

As pointed previously, the color shades of the dyed samples can be affected by the mordant type used. The alum and tin yield beige and light brown color; chrome offers dark yellow to light greenish brown; iron shows reddish brown to greenish brown, and copper gives yellowish green to reddish brown.

Pinecone produces a brown and reddish yellow on wool. The entire mordanted samples dyed using pinecone fall within colour space of CIELab in the range of red-yellow. The dyed wool without mordant and the dyed wool with tin are observed to exhibit the greater redness and lesser hue angle. Increase of the pinecone percent from 25 to 100 on fibre weight clearly shows the relative shifting of the brown shade of pinecone towards more reddish brown colour as pointed by FTIR studies. Nearly similar coordinates are seen in the raw wool, the dyed un-mordanted wool and the dyed Cr-wool. In the case of dyed wool the differences in colour coordinates using different mordants are found to be pertinent. A great difference arises in the colour coordinates relating to raw and dyed wool specimens, whereas in the case of specimen dyed with mordant the two colour coordinates fall in nearly same range.

The absorbance of residual dye in dyebath having copper and ferrous is considerably lesser in comparison with other types of mordants. The absorbance of solution directly relates to teh concentration of colour bath (as per Beer's law) the dye uptake of wool that is pre-mordanted using iron and copper is higher than other mordanted fibres [53]. Also, the reduction of absorbance of remaining dye in dyebath leads to increase of the exhaustion of dye bath. Hence, in the case of pre-mordanted wool using copper and ferrous the values of low absorbance of residual dye exhibit high percentage of exhaustion of dyebath.

3.3   Differences in Color

Considering the coordinate space the difference between two colours is denoted by Delta E (ΔE). A rating value between 3.5 and 5 indicates that the difference in colour between the specimens is good and clear. The difference in colour is brilliant at value more than 6 [53]. Differences in colour pertaining to dyed wool which is unmordanted have been determined. Based on values of Delta E, values above 8 arise from the difference between raw and dyed un-mordanted wool. The findings reveal that the rise in the concentration of dye has a direct bearing on values of ΔE. Also, the differences in colour between dyed samples with percentages of 25, 50 and 100 on fibre weight of dye become considerable. Hence, improving the shade depth of color is rendered effective by the pinecone concentration.

3.4   Color Strength (K/S)

The values of colour strength also change significantly by the various mordants besides creating differences in hue color and CIE lab values L. The increase of dye concentration enhances the color absorption. Because of optimum absorbance and formation of metal complexes with wool its color absorption gets severely affected by the mordants used [53]. By the use of copper and ferrous sulphate deep shades are achieved in the mordanting process respectively. The lowest K/S value is achieved with stannous. Iron and stannous show the maximum and minimum values of colour strength in wool dyeing wool using barks of *Ficusreligiosa* respectively.

3.5   Characteristics Relating to Color Fastness

Tests for wash and light fastness relating to mordanted and un-mordanted wool dyed with pinecone have been done and their values determined. In the case of un-mordanted wool the color fastness results show that the wash fastness is rated well lying between 4 to 5 (with maximum rating of 5). The attraction of wool can contribute such finding to a pinecone. The light fastness of un-mordanted wool is to be medium (3–4 on the standard blue scale, whereas 8 is the highest rating) [53]. A good value of light fastness of 5–7 is observed in the samples dyed by pre-mordanting method (8 is the highest rating). There is an improvement in the light fastness by a grade of ½ to 1 with the increase in the dye concentration. Copper or ferrous sulfate imparts high resistance to fading, but it is not the case with stannous chloride or alum. The ligands in the pine cone like OH and COOH can contribute to the high fastness of this dyed sample, and form a complex with a metal ion from the mordant.

4   Use of Kum Leaves on Various Dyeing Mordants

Natural dyes have a very old history as being part of the human life. In the areas of craft and academics, natural dyes are making a comeback. The synthetic dyes can give strong colours, but are carcinogenic and inhibit benthic photosynthesis [54]. Even though natural dyes have been used to a less extent during the past few decades, their use have never been stopped altogether and are still being used in various areas around the globe [55].

The art of natural dyeing have originated in India. It still continues to be common in India. Manipur in india is well known for its handlooms and handicrafts. The *Meitei* communities have been using the mature leaves and young shoots of *Strobilanthes cusia* (Nees) Kuntze (*Kum* in Manipuri) for producing a unique blue-black and indigo coloured *Kum* dye. *Kum* dye is used for dyeing *Kum phanek*, a formal dress worn by Meitei women woven on loin loom. The dyeing of *Kum* is an indigenous dyeing technology which has come down generations. The knowledge about the extraction and dyeing methods dates back to the 11th century A.D. and still persists in a few valley localities [56, 57]. *Kum* is the most important plant used for dyeing fabrics by different communities in Manipur [58–64].

Owing to lack of proper maintenance of records the use of natural dyes has stopped over the years despite aboriginal system of knowledge being practiced for many years. Effort has been taken to restore the ancient natural method of dyeing. The mature leaves and young shoots of *Kum* have been fermented and their influences on the dyeing process variables investigated. Assessment of colour fastness properties (washing and rubbing) of dyed cotton, silk and woollen fabrics at optimal conditions, with various doses of natural as well as chemical mordants have been done. The tensile strength and elongation percentage of the grey as well as dyed yarn have also been determined.

S. Cusia of the family Acanthaceae is a wild perennial shrub growing to 13–18 cm, leaves are 10–18 cm long, membranous with 6–7 lateral nerves on either half, elliptic, ovate and acute at both ends. Flowers are purple in densely panicled spikes, usually opposite with ovate, deciduous bracts. Calyx segments are linear-spathulate and corollas are 4 cm long and glabrous [65].

The fermented *Kum* when dyed on silk and woolen yarn shows good results in relation to colour fastness to washing and rubbing, tensile strength and elongation percentage and prove beneficial to textile industries. Owing to the mordant used all the dyed yarns show good fastness properties. The cumbersome and long dyeing process in traditional method of cold dyeing leads to poor fastness property of yarns. Hence it can be replaced by heating process following the traditional method.

4.1   Influence of Metallic Salts

The colour evaluation pertaining to rub fastness and wash fastness, tensile strength and elongation % of all the pre-mordanted samples with aluminium sulphate, copper sulphate, ferrous sulphate and potassium dichromate have been determined. The results pertaining to colour fastness to rubbing in dry and wet grades and wash fastness in cotton and woollen yarns range between medium to good (3–4) in all the pre-mordanted samples. The colour fastness to rubbing and washing for almost the treated samples in silk exhibit excellent to good (4–5) results. There is no colour staining to adjacent cotton fabrics (4–5).

With regard to cotton and woollen yarns, in the mordanting method using metallic salts the tensile strength is higher than that in grey yarn. The tensile strength of 2% $FeSO_4$ treated silk sample is lower in comparison to that of grey yarn. Elongation percentage in dyed woollen yarn is higher than in grey yarn but there is less elongation % in cotton yarn. In 10% $Al_2(SO_4)_3$ and 2% $CuSO_4$ treated silk samples the elongation % is higher than in grey yarn but it is lesser in 2% $FeSO_4$ and 2% $K_2Cr_2O_7$ treated samples.

4.2   Influence of Natural Mordants

Medium to good results (3–4) are obtained in traditional hot dyeing method. Good to excellent (4–5) result is achieved in silk yarn. In the case of cotton and woolen yarn poor result is seen with rub fastness in wet grade. However, medium to good result (3–4) is seen in the overall rub fastness in dry grade. Greater tensile strength is observed in dyed cotton and woollen yarn in three different ratios of mordant than that of grey yarn. In dyed silk yarn the value is lesser than in grey yarn for 2:3:3 and 3:3:2 ratio treatment. Elongation % of dyed cotton yarn is higher than that of grey yarn for 5:2:1 and 2:3:3 ratios and it is lesser than in grey silk and woollen yarns.

In traditional cold method, the rub fastness (dry and wet) exhibits poor to medium (2–3) result in cotton yarn and fair to good (3–4) result in wash fastness. The colour fastness to rubbing as well as washing, exhibits fair to good (3–4) result.

It has been found that as compared to the grey yarn the tensile strength of all the dyed yarns shows greater values. In case of elongation %, the cotton yarn has almost high value, and the silk yarn has lesser elongation break than the grey yarn. In comparison with grey yarn the woolen yarn exhibits lower value of elongation break except for 3:3:2 treatment.

When Acalypha and related natural dyes have been applied on silk rub fastness in wet and dry states has been reported to be good. In the areas of cutch and ratanjot rub fastness in dry state has been found to be good. However, average wet rub fastness has been noticed [66, 67]. As majority of the natural dyes are highly soluble in water and hence show poor colour fastness to washing. In order to improve its colour fastness property the mordants are necessarily be used [68]. The attraction between the dye and fabric have been mordants which are metal salts [68, 69]. Most mordants are mineral salts, the most common being aluminium, iron, copper, tin and chrome. The effects of various mordants have a crucial function in causing fading of eighteen natural dyes that are yellow coloured [70]. Mordanting technology improves the development of shade and provides a link to colouring substrate for fixing on cloth.

In all such cases medium to good rub fastness properties show that there are almost entire natural dyed that are not fixed and remain superficial on surface of fibre get almost removed by means of soaping and washing treatments. The natural dye is able to enter inside the openings of the fibre and could possibly be fixed well through co-ordinated complex formation with the mordants and mordanting assistant.

Colour fastness to washing and rubbing for application of chosen dye at greater alkaline *p*H, with/without treatment with dye fixing agents is always found to be better as greater alkali concentration leads to better ionization of dye anion and anion formation (preventing aggregation of dye molecule) to take part in complex formation among the dyes and mordants. The alkaline *p*H of the dye bath enhances the colour fastness of jute dyed using red sandalwood [71]. There are even changes in thermal transitions of some natural dyes when subjected to heating (equivalent to dye bath temperature or higher). However, both light fastness and wash fastness depend on the nature of pre-treatment (alkaline/acidic) on the fabric before dyeing [72].

The extension of break i.e. the elongation % for the treated samples is higher than the untreated one due to the fact that the influence of mordanting with zirconium salt leads to produce more elastic wool fibres [73].

The stress of sewing is withstood due to the high tensile strength of the yarns. The strength of silk coupled with elongation determines toughness of materials which relates to weavability [74]. The high tensile strength in silk yarn can be attributed to its proteinaceous nature of the fibre which gets easily bound with metal ions present in mordants. Increase in mechanical strength is due to the application of metal ions that improve fibre resistance. The influence of metal ions on silk varies with the type of mordant used and amount absorbed [75]. However, there is reduction in value of tenacity after dyeing. It arises due to chemical reactions and heat treatment during pre-treatment and dyeing processes, which causes reduction in the reduction of the degree of polymerization of cotton fibre. This is so despite the strength of cotton fibre being greater in wet state than in dry state [76]. Owing to absorption of dye molecules the peptide bonds or salt linkages have strong inter polymer forces of attraction and hence contribute to the cohesion of appropriate fibre polymer system. It subsequently improves tenacity, elastic nature and durability of textile fibre and thus textile material [77].

There is an improvement in single yarn strength in mordanted specimen in comparison with undyed non-mordanted yarn. The increase in breaking strength can result from the increase in the size of dye molecule after using mordants, which penetrates into the fibre core and might have, in turn, increased the strength [78]. With the increase in concentration of chitosan, there is increase in the breaking strength values of cotton yarns. With cotton yarns, there is reduction in elongation at break of the yarn with the increase in the breaking strength values. In the case of chitosan the resistance to axial load is improved with such binding of the fibres in yarn [79].

5   Dyeing with Natural Dye Extract Using Synthetic and Natural Mordants

A wide range of rich colours that are complimentary to one another arises from natural dyes [80]. A number of compounds are present in natural dyes of plant origin and differences in their properties arise from the silk kind and weather conditions. There has been an abrupt decrease in the use of natural dyes for more than 150 years, with increased use of synthetic dyes, since existing natural dyes have been unable to satisfy the demand of the market. Textile dyeing with natural dyes is backed by a rich tradition in certain parts of India. Natural dyes continue to be applied on traditional wool and woolen products at many places of India. The dyeing of textile materials with natural dyes have innumerable drawbacks relating to properties and processes [81].

The attraction between dye and fabric is created by mordants which are metal salts [82]. The widely used mordants are alum, chrome, stannous chloride, copper sulphate and ferrous sulphate. With a given type of dye material many shades and tints are produce by natural dyes [83]. The fastness properties with regard to grey scale of *Symplocos racemosa* species (from which natural dyes are derived) is good. So far there is no literature reporting isolation of natural dyes from this species [84, 85]. The ancient practice of dyeing using natural dyes has been revived by research studies [86]. Optimum dyeing conditions for wool have been adopted with the leaves extract of *Symplocos racemosa*, with combination of mordants. Parameters such as washing, rubbing, perspiration and light fastness relating to colour fastness have been evaluated [87, 88]. The capability of a material to withstand any change in its colour properties or degree of shift of its colourants to neighbouring contact white materials is termed as color fastness.

The extracted dye from the leaves of *Symplocos racemosa* can be effectively used in wool dyeing for achieving many light and soft colors through mixing of natural and synthetic mordants. The test samples show excellent colour fastness to washing and rubbing (excluding combination for pre-mordanting with lemon juice: potassium dichromate and combination of lemon juice: stannous chloride), ranging between good to excellent for perspiration fastness in acidic and alkaline media and fairly good light fastness. The textile sectors will derive benefit from this information.

5.1   Lemon Juice: Stannous Chloride

The colour fastness to light, washing, perspiration and rubbing have been determined in the case of dyed wool samples treated with combination of lemon juice: stannous chloride in water medium. In all the proportions of the mordant combinations studied the treated wool samples exposed to light exhibit good fastness to light. The grades for wash fastness in the treated samples prior to mordanting have been found to be average. However, the wash fastness have been found to be good to excellent for the treated wool samples in the case of simultaneous and post mordanting [89].

No color staining is noticed. The change in colour to rubbing in dry and wet states has been rated as excellent for all the treated samples. Colour staining has not been noticed in the range of no staining and negligible staining in the case of dry rubbing. The grades for fastness to perspiration have been found to be good to excellent, except in the case of mordant ratio of 3:1 for pre-mordanting technique, in which case it is found to be average, for all samples in acidic as well as alkaline media. In the case of all the samples treated in acidic as well as alkaline media no colour staining has been observed (5).

5.2   Lemon Juice: Copper Sulphate

The assessment has been done for evaluation of colour fastness to light, washing, rubbing and perspiration of dyed wool samples treated with lemon juice: copper sulphate combination in aqueous medium. Nearly all the treated samples subjected to light exhibit fairly good (4) light fastness for all ratio mordant combinations. The treated samples for pre mordanting exhibit fair (4–5) wash fastness grades, but they range between excellent and good (4–5) in the case of all the treated samples for simultaneous and post mordanting [89]. Color staining has not been observed. All the treated samples show the colour change due to dry and wet rubbing, which is found to be excellent (5). With dry rubbing, there has been no colour staining that ranged between no staining and negligible staining (4–5). The perspiration fastness grades range between 4–5, except for 1:3 the proportion in pre-mordanting method, where it is fair (4), for all samples in both acidic and alkaline media. There is no colour staining (5) for all the treated samples in both acidic and alkaline media.

5.3 Lemon Juice: Potassium Dichromate

Evaluation has been done with regard to colour fastness to light, washing, rubbing an perspiration for dyed wool samples treated with combination of lemon juice:potassium dichromate in aqueous medium. Almost the entire treated samples exposed to light show reasonably good fastness to light for mordant combinations of all ratios. In the case of all treated samples the grades of fastness to washing have been found to be reasonably good. However, for mordant ratio of 1:3 in pre-mordanting method, it is found to be average. The fastness to colour against rubbing in dry and wet states has been found to be excellent for all the treated samples. In the case of dry rubbing the colour staining there is no staining and there is negligible staining in the case of wet rubbing. But, in the case of pre-mordanting technique it is found to be average [89]. The fastness grade to change of colour is found to be excellent for most treated samples. However, it is found to be good in the case of mordant ratio of 1:3 in pre mordanting techniques. Under alkaline as well as acidic medium no staining of colour has been observed in the treated samples. In the case of post and simultaneous mordanting techniques, significant results have not been noticed in respect of excellence in the fastness properties.

5.4 Lemon Juice: Ferrous Sulphate

The assessment with regard to fastness of light, rubbing, washing and perspiration have been has been determined in the case of dyed wool specimens treated with combination of lemon juice: ferrous sulphate in aqueous solution. For mordant combinations of all proportions the treated specimens exposed to light exhibit reasonably good fastness to light. In the case of all treated wool specimens the grades of fastness to washing fall between good to excellent. The treated wool specimens show excellent fastness of colour to rubbing in dry and wet states. In the case of rubbing in dry state the staining in colour is found to be almost average. The grade of change in colour is excellent for nearly all treated wool samples. However, under alkaline as well as acidic conditions it is found to be good for mordant ratio of 1:3 in simultaneous mordanting. Under acidic as well as alkaline conditions there is no staining of colour for the treated specimens. Extracted natural dye from leaves of *Symplocos racemosa* shows yellow colour. Use of various techniques of mordanting helps to achieve a wide range of shades of colour [90–93]. When various mordants like $K_2Cr_2O_7$, $CuSO_4$, $SnCl_2$ and $FeSO_4$ are used different colour shades are obtained. Generally as synthetic or chemical mordants, $K_2Cr_2O_7$ gives pale yellow colour, $CuSO_4$ shows light green colour, $FeSO_4$ gives brown colour and $SnCl_2$ shows light yellow or cream colour with dyes on wool fibres [89]. The mordanting of wool with different concentrations enables achievement of a number of shades. The various metallic mordants yield various shades of colours than those of all natural dyed samples.

6   Kinetics and Thermodynamics of Natural Dyeing with Herbal Extract

In India, *Arnebia nobilis* Rech.f., also termed as 'Ratanjot', has traditionally been a crucial natural source of red colour in the area of pharmaceuticals, cosmetics and food [94, 95]. Studies have already been conducted to isolate and identify the components of *Arnebia nobilis* Rech.f. Alkannin β, β-dimethylacrylate has been identified as the major component constituting ~25% of the crude extract [96].

Earlier work reported the results of the influence of pH and temperature on the crude dye [97]. A high sensitivity to pH is seen in the dye. The dye has also been found to be heat sensitive and exhibits degradation beyond 80 °C. Textile substrates of different types like nylon, polyester, acrylic, wool, silk and cotton, have been dyed at various pH values. All substrates showed good affinity with nylon dyed in blue, polyester dyed in pink and other substrates in purple colour at pH 4.5. According to Indrayan et al. and Kyu and Soo, the active coloured ingredient in *Arnebia nobilis* is present in quinonoid form in acidic medium [98, 99]. In alkaline medium, the phenolic proton of quinonoid form gets dissociated from the naphthoquinone nucleus and is converted to benzenoid form which is responsible for blue colour.

Earlier investigations have been carried out on natural dyes having various chemical structures in order to understand the mechanism of dyeing synthetic fibres. Some elaborate investigations have been conducted out on quinonoid dyes. In the dyeing of synthetic fibres such as nylon and polyester two isomeric dyes derived from naphthoquinone, namely, lawsone and juglone (2-hydroxyl and 5-hydroxy naphthoquinones respectively) have been found to yield linear isotherms. Both the dyes showed very high affinity towards hydrophobic fibres [100, 101]. Certain studies have focused on the naphthoquinones of *Onosma echioides*, also known as 'Ratanjot' in literature. The dye has been found to be adsorbed by nylon as well as polyester in the similar way as a disperse dye [102, 103].

Good affinity towards nylon and polyester has been shown by the natural anthraquinone-based colourants extracted from madder roots. The mechanism of dyeing conforms to the Nernst isotherm [104, 105]. But, varying results have been achieved when nylon and polyester have been dyed with long, conjugated carotenoid molecule Bixin. The dyeing of nylon conforms to Langmuir isotherm while dyeing of polyester conforms to Nernst isotherm [106]. Berberine (C I Natural Yellow 18), a natural basic dye, was sorbed by acrylic by site mechanism indicated by Langmuir isotherm [107]. The dye extracted from *Rheum emodi* has been used to dye wool and silk. The dyeing corresponded to partition mechanism, confirming the anthraquinonoid structure [108]. There has been a good affinity towards nylon and wool fibres by the dye extracted from red sandalwood and the mechanism conforms to partition type similar to that of disperse dye on hydrophobic fibres [109].

It has been noticed that most of the naphthoquinone and anthraquinone based natural dyes investigated have shown high substantivity for synthetic fibres. The dyes have been found to behave as disperse dyes and have shown partition mechanism on synthetic substrates. Little work has been found in literature relating to explanation of the mechanism of dyeing natural fibres with disperse dyes. Since earlier investigations have shown positive results on wool, it provided an interesting opportunity to carry out kinetic and thermodynamic investigations of crude dye extracted from *Arnebia nobilis* Rech.f. on wool to understand the theoretical basis of dyeing. The diffusion coefficients of the dye has been compared with those of the other naphthoquinone natural dyes.

The dye extracted from *A. nobilis* Rech.f. exhibits good affinity for wool fabric. There is a close relationship between the mechanism of dyeing mechanism and the partition mechanism, which confirms that the naphthoquinonoid based dye acts as a disperse dye on wool. Heat of dyeing is found to be negative and the dyeing process appears to be exothermic. The entropy is also found to be negative.

6.1   Investigation on Kinetics

6.1.1   Dyeing Rate

For wool fabric dyed at 80 °C the dye uptake for different time durations has been determined. The half dyeing duration has been calculated to be 60 min. It has been found that as the dyeing duration increases from 15 min to 600 min (10 h), there is increase of dye uptake from 10 g of dye/100 g of fabric to 80 g of dye/100 g of fabric. A further increase in time to 1500 min (25 h) leads to a slight increase in dye uptake of about 90 g of dye/100 g of fabric. However, no significant dye uptake is observed thereafter, as the dyeing time approaches infinity i.e. 2880 min (48 h), thus indicating that the equilibrium is achieved.

It has been reported by Das et al.[15] that the non-polar hydrocarbons of the protein fibres are considered to be chiefly responsible for the incorporation of the non-polar dye in the structure of wool.

6.2 Diffusion Coefficient

Based on the Hill's equation, Urbanik approximation, and Rais and Militky approximation, the apparent diffusion coefficient of the crude dye of *A. nobilis* is found to be 0.25×10^{-11} cm^2/s, 0.22×10^{-11} cm^2/s and 0.21×10^{-11} cm^2/s respectively. All the equations are found to yield comparable values.

Comparison has been done on the D_{app} values obtained for crude extract of *A. nobilis* and other naturally occurring naphthoquinone and anthraquinone dyes on wool. The diffusion coefficients of all such dyes have been determined. The D_{app} value of the dye ($\sim 0.2 \times 10^{-11}$ cm^2/s) is found to be comparable to that of *Rheum emodi* (0.2×10^{-11} cm^2/s). However, it is found to be much lower than that of Juglone and Lawsone which has D_{app} values of 6.08×10^{-11} and 2.58×10^{-11} cm^2/s respectively.

On comparing the diffusion coefficient of the crude dye extracted from *A. nobilis* on wool with those on other synthetic substrates dyed with the same dye, the diffusion coefficient of the dye on wool is observed to be much less than that on nylon and polyester (56.4×10^{-11} cm^2/s and 1.86×10^{-11} cm^2/s respectively). It may be because the rate of diffusion increases as temperature increases. Thus, greater diffusion coefficient is achieved on nylon and polyester that have been dyed at 90 °C and 130 °C respectively, than wool that has been dyed at 80 °C.

6.3 Adsorption Isotherm

A quantitative analysis of the dye in fabric as well as that in dyebath has been carried out and results are plotted as adsorption isotherm. For prediction of the nature of the isotherm, the best fit line for three models of dye sorption viz. Nernst, Langmuir and Freundlich has been drawn, and has been used to define the theoretical model for a specific system of dyeing. Hence the model provides a base for the calculation of thermodynamic factors. The dye molecule of *A. nobilis* is supposed to be very small and simple without presence of ionic groups. Such are the properties of a disperse dye and, hence, theoretically the isotherms should conform to the linear or partition mechanism of dyeing.

The isotherm of wool fabric is depicted in Fig. 2. The best fit isotherm is linear, having a high correlation coefficient ($R^2 = 0.967 - 0.973$), that implies the partition mechanism of dyeing or Nernst model. In the case of dyeing with disperse dyes on synthetic fibres such model is usually observed. Linear isotherms have also been achieved in dyeing of natural protein fibres with anthraquinone dyes and in dyeing of wool with juglone and lawsone. The dye uptake at equilibrium is found to be highest at 80 °C. The slope of the isotherm, which is indicative of partition ratio, increases from 80.72 to 119.6 with the increase in dyeing temperature between 70 to 80 °C. However, it is observed that the slope of the adsorption isotherm decreases to 69.57 with the increase in dyeing temperature to 90 °C. This may be due to the

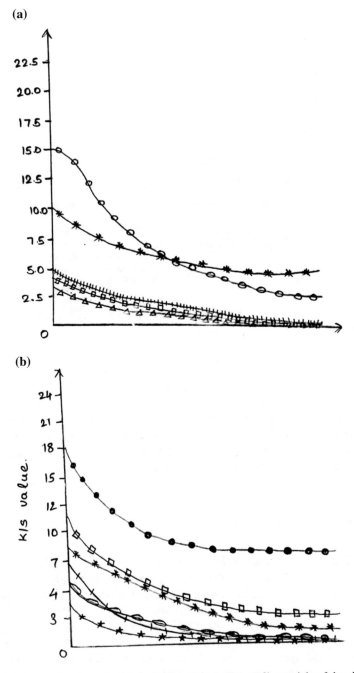

Fig. 2 Color strength values of wool dyed using [53]. **a** 25% on fibre weight of dye. **b** 50% on fibre weight of dye. **c** 100% on fibre weight of dye

N. Gokarneshan

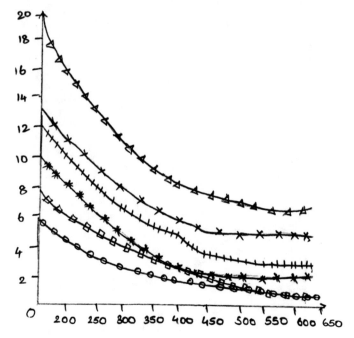

Fig. 2 (continued)

decomposition of the dye molecule at higher temperature. The slopes and results of statistical analysis for best-fit isotherms have been determined.

During dyeing, it is observed that a large amount of dye is being taken up by the fabric at higher concentrations of dye liquor and during longer hours of dyeing. Fabrics which are red during initial hours of dyeing become brownish black to black when dyed up to the equilibrium with infinite solutions. Such shade build-up can arise from the aggregation of dye inside the fibre.

6.4 Thermodynamic Parameters

Dyeing of wool with *Arnebia nobilis* corresponds to the partition mechanism, and hence the standard affinity of the dye for wool fabric has been calculated.

The standard affinity values have been achieved at 70, 80 and 90 °C. The value of standard affinity is found to increase from 15.92 to 17.8 kJ/mol as the dyeing temperature is increased between 70 to 80 °C. A further increase in temperature by 10 °C results in the reduction in standard affinity to 15.48 kJ/mol. Such trend is also clear from the adsorption isotherms. The affinity of many natural dyes on wool dyed at temperatures varying between 80 and 100 °C has been found to vary from 7 to 29 kJ/mol. An affinity of 17.8 kJ/mol of *A. nobilis* at 80 °C is found to be comparable.

6.5 Heat of Dyeing

Heat of dyeing has been calculated considering the temperatures of dyeing at 70 and 90 °C; and the standard affinity at temperatures 70 and 90 °C. It is found that for dyeing of wool with this dye, the heat of dyeing is negative (-23.48 kJ/mol) (exothermic dyeing process) and hence, with rise in temperature the dye absorbed at equilibrium will be low [109]. The dyeing process is found to be endothermic on comparing the heat of dyeing of *Onosma echioides* (Ratanjot) on synthetic substrates (nylon and polyester).

The heat of dyeing is determined by the force of bonding between the fibre and the dye. The enthalpy of heat (H°) is shown as the energy of bonds that are broken and formed. A more stable dyeing result from increase in value of energy of bonds formed with dye and fibre. The heat of dyeing becomes negative when the value exceeds the energy of broken bonds [110]. Hence, more bonds are formed leading to negative heat of dyeing of *A. nobilis* on wool.

6.6 Dyeing Entropy of Dyeing

The third thermodynamic parameter, entropy of dyeing has been calculated. The calculated value of entropy is -22.05 J/mol/K, and the dyeing entropy is also found to be negative. The negative entropy is attributed to a uniformly order distribution of dye in the substrate. An orderly arrangement of dye molecules along fibre axis as the dye molecules are absorbed by the fibre. A negative entropy arises since dye molecules can move less with a small possibility leading to decreased entropy. However, entropy of nylon and polyester dyed with *Onosma echioides* is found to be positive.

7 Keratin as a Cross Linking Agent in Dyeing

Keratin is a protein that exists in the external skin layer of human beings as well as many other creatures. It is tough, and insoluble by nature and is related to a group of structural proteins richly present in wool, hair, feather, hooves, and fingernails. Such keratinous materials are often known as "hard" keratins (in contrast with soft keratins which are present in epithelial tissues).

Even though there are huge quantities of keratin wastes, keratins also find some uses in the global market, of which feather meals are an example used for animals [111, 112]. Keratin-based cosmetics are used for treatment of human hair and skin [113, 114]. Keratin materials have made their entry into many other areas, such as concrete, ceramic, fertilizers, fire fighting compositions, wound healing, leather tanning, production of biohydrogen, and shrink-proofing of wool [115–122].

A number of hydrolytic and non-hydrolytic techniques have been used to extract keratin from natural resources. Most of such techniques use oxidizing or reducing agents in presence of auxiliaries like sodium dodecyl sulphate, urea, and EDTA [123–125].

A simple technique has been used to extract keratin from cheap coarse Egyptian wool fibres. As the extracted keratin is a natural proteinic biopolymer, its dyeability can be enhanced with reactive as well as acid dyed by crosslinking it with wool. In order to achieve this wool has been treated with keratin/ECH mixture, and thus reduce the amount of dye in the effluent. The influence of various dyeing conditions on the dyeability of the treated and untreated wool fabrics has been investigated.

Epichlorohydrine can be used to permanently cross link the biopolymer keratin to wool. There has been improved dyeability in the treated wool using mono-sulphonic and disulphonic acid and vinyl sulphone and α-bromo acrylamide reactive dyes. As keratin is a biodegradable protein it can be an acceptable option for other unacceptable chemicals used traditionally in wool chemical processing. More investigations are being carried out to apply keratin to certain synthetic fibres to achieve specific desired dyeing properties.

7.1 Influence of Dyeing Temperature

Untreated and treated wool fabrics have been dyed with C.I. Acid Blue 203, C.I. Acid Red 1, C.I. Reactive Blue 69, and C.I. Reactive Red 180 using 4% shade, 5 pH, 60 min treatment time, 1:100 MLR and different dyeing temperatures (25°–85 °C) [126].

As can be seen from the results (Fig. 3) that in case of untreated and treated fabrics, the extent of dyeing (K/S) increases with temperature of dyeing, immaterial of the dye used [126]. It has been found that the colour strength value of the treated fabrics is a little higher than untreated wool at room temperature (30 °C). But the shades of these samples are still too light and attained K/S of 5–10. The highest K/S value is only achieved at 85 °C for 60 min in dyeing with reactive dyes for both treated and untreated wool.

7.2 Influence of Wool Treatment on Dyeability

As keratin macromolecules are proteinic in nature, they should bind chemically to wool. Epichlorohydrin is included as a crosslinker so as to ensure permanent crosslinking of keratin onto wool. Investigation has been carried out regarding the influence of treatment of wool using keratin/ECH on its dyeability with the acid dye Supranol Blue BLW (C.I. Acid Blue 203) at various temperatures (40°–85 °C) [126]. The dye bath used in dyeing of pretreated wool test fabrics using Acid Blue 203 is nearly completely exhausted within 5–30 min, based on the temperature of

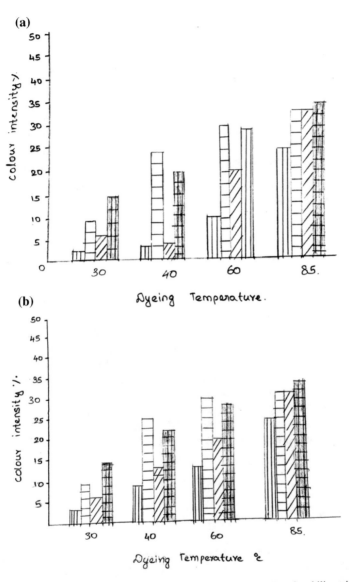

Fig. 3 Influence on wool fabrics treated with keratin/ECH with regard to dyeability with acid as well as reactive dyes [126]

dyeing. This time is much shorter than the time needed for complete exhaustion of the dyebath in case of dyeing the untreated wool fabric using the same dye. The dye bath gets fully exhausted within half an hour by pretreatment of wool fabric upon dyeing at room temperature, or within 5 min upon dyeing at 60 °C.

Such findings can be describe with regard to measurement of the ability of keratin to form permanent crosslinks with wool polypeptide chains. The modified wool has greater number of cationic groups which results in enhanced absorbability of the anionic acid dye. Being protein containing a lot of amino and amide groups, in acid medium, keratin has sufficient cationized basic nitrogen-containing groups able to bind with the anionic sulphonic groups in the acid dye, as shown below:

$$HOOC-W-NH_2 + Cl-ECH-OH + HOOC-K-NH_2 = (Wool)(Crosslinker)(Keratin)$$
$$HOOC-W-NH-ECH-OOC-K-NH_2$$

7.3 Kinetics of Dyeing

The values of half-dyeing time, specific dyeing rate constant and the apparent diffusion coefficient calculated for the untreated and keratin-treated wool fabrics dyed with C.I. Acid Blue 203 are obtained. The obtained data show that, immaterial of the dyeing temperature and the dye yield, the half-dyeing time of the keratin-treated wool fabrics reduces sharply compared to the untreated one [126]. Again, the enhanced rate of dyeing might be attributed to the extra cationized amino and amide groups created along the polypeptide chains of wool as a results of its chemical bonding with keratin. There is an appreciable increase in the specific dyeing rate constant (K') of wool fabrics when treated with keratin/ECH. For dyed wool fabrics treated with keratin the diffusion coefficient (D) is found to be higher than that of the untreated one.

7.4 Fastness Properties

The washing and crocking fastness of untreated and keratin-treated wool fabrics dyed with C.I Acid blue 203 have been determined [126]. The fastness properties of wool fabrics against washing and crocking are not affected after treatment with keratin/ECH. This indicates that keratin is bound to the wool fabrics by permanent cross-links which are fast to washing and rubbing conditions.

7.5 *Scanning Electron Microscopy*

Scanning electron microscopic investigation has been carried out for untreated and keratin treated wool fabrics so as to monitor any change in the morphological structure of wool fibre after being treated with keratin/ECH. The normal surface morphology of the untreated wool fibres wool fabric treated with keratin/ECH results in formation of a very thin layer of cross-linked keratin on the surface of wool fibres [126]. Nevertheless, part of the added keratin, cross-linked with the fibre interior leads to durable dyed wool.

This criterion is confirmed by measuring the solubility of untreated as well as treated wool fibres in urea bisulphite solution. The urea-keratin ECH-e bisulphite solubility of the treated wool fabric is found to be 4.6%, compared to 39.4% for the keratin treated, and 44% for the un treated one. This is a strong clue that keratin in presence of epichlorohydrin forms cross links non only with the fibre surface but also with the different chanses of the bulk of the fibre. This hypothesis is in harmony with the results of the fastness properties of the dyed wool fabrics.

8 Influence of Bio Carbonization

Raw wool contains varying percentages of vegetable matters as impurities depending on the sheep breed [127]. The vegetable contaminants in raw wool fibres accounts up to 8.5%. It indicates that effective mechanical method of elimination of impurities is difficult and hence necessitates more severe chemical treatment [128, 129].

During wool carbonization the vegetable matters and skin flakes are removed. In order to convert the cellulose into easily removable hydrocellulose sulphuric acid which is a strong acid, is used. In order to reduce the damage to wool fibre, effluent waste and consumption of energy replacement of chemical carbonization by enzymes has been studied [130, 131]. For separation of wool from vegetable impurities Bio-carbo process is used and involves biologically active agents comprising of certain enzymes with low quantity of specific chemicals [132]. In the same manner, elimination of burr is rendered easier after incubating wool with cellulases, burr removal became because of weakening of cohesion between wool and burrs [133]. Upon the elimination of the enzyme treated burrs by mechanical means it is observed that wool does not suffer physical or chemical damage.

In treating raw Egyptian wool attempt has been made to utilize certain enzymes obtained from market. The effect of bio-carbonization treatment of wool has been studied on its dyeability with acid and metal complex dyestuffs in equal proportion.

A mixture of cellulose, pectinase and xylanase enzymes has been used to effectively purify the Egyptian wool fleece from vegetable matters while taking care to preserve the inherent properties of wool. There is no loss in fibre strength and weight as the enzymes are very particular hydrolyzing agents. These enzymes can be combined with mechanical cleaning techniques or milder chemical techniques in the case wool fibres that are contaminated strongly.

8.1 Influence of Treatment with Single Enzyme

The elimination of natural impurities from raw wool has been effected by the use of chosen cellulose, pectin and lignin digesting enzymes. The technique has been compared with the traditional carbonization process that uses dilute sulphuric acid [134]. When raw wool fibres are treated with sulphuric acid it leads to nearly entire elimination of the natural impurities from wool. But, during the carbonization process the tensile strength of the carbonized fibre reduces to approximately 25%, in relation to raw wool fibres possibly because of partial hydrolysis of the peptide bond. The carbonization of wool with sulphuric acid has been found to lead to extensive fibre breakage in subsequent processes. Improper rinsing and neutralization of wool is found to cause some further damage.

When the carbonization of wool fibres takes place by use of sulphuric acid it results in a restricted increase in their alkali solubility. It can be because of the partial reaction of sulphuric acid with wool due to influence of the drying states. The extent of increase of solubility of alkali can arise from formation of sulphonic or sulphate groups. Based on the amount and nature of enzyme utilized the vegetable matter/natural impurities is separated from wool to different degrees by treatment of raw scoured wool fibres using cellulase, acid pectinase or xylanase. The vegetable matter that remains in wool is about 25% of the total impurity content when 20 mL/L Biotouch L is used. It is, possibly because of the particular types of enzyme used. The vegetable matter ia acted upon by Biotouch L, while the pectin is destroyed by acid pectinase, and lignin is separated by xylanase. Since pectinase or cellulose enzyme has been applied on wool fibres in concentrations which are high, there has been a little increase in the tensile strength.

The solubility of the alkali of the fibres treated with enzyme does not considerably alter than in the case of raw. Again, it can be because of the use of particular enzymes that cause separation of natural impurities from wool keratin. This proves the merit of enzymes as compared with sulphuric acid.

8.2 Influence of Mix Enzyme Treatment

Scoured wool fibres have been treated using the mentioned enzymes under prescribed temperature and pH. The influence wool treatment with this mixture of enzymes has been determined. Compared to the single or double enzyme system the combined action of the three enzymes in one bath is far more effective for separation of natural impurities from raw wool. For raw wool the % of natural impurities shows decrease from 4 to 0.16 for wool pretreated with combination of 3. It can be because of the fact that each enzyme is capable of causing separation of a certain portion of the natural impurities which cause contamination of raw wool.

8.3 Influence of Duration of Treatment

A combination of three enzymes have been used to treat raw wool for different time spans so as to reduce the treatment duration and thus the cost of the bio carbonization process for wool. When wool fibres have been carbonized using sulphuric acid a restricted increase in their alkali solubility occurs as the treatment time increases. It can be because of the partial reaction of sulphuric acid with wool by means of the influence of the drying states. It can result in formation of sulphonic or sulphate groups that enhance the degree of solubilisation of alkali.

Raw wool has been treated with combination of 3 enzymes for various time durations in order to reduce the duration of treatment and thus the cost of the bio-carbonization process for wool. For enzyme treatment of wool, increase in treatment time till 2 h, there is reduction in the impurities content in the treated fibres. There is no considerable influence on the degree of elimination of natural impurities and the tensile strength and alkali solubility of the treated wool with further increase in the duration of treatment.

8.4 Properties of Dyeing

Wool fibres not treated as well as pretreated with 3 enzymes have been dyed with selected (acid/metal complex) dye in the proportion of 1:1 between 85 and 95 °C separately for different time durations and the percentages of exhaustion of dye studied. The wool scouring has resulted in appreciable improvement in the dye exhaustion with metal complex/acid dye at the two temperatures specified. It can be because of the elimination of greasy matters from wool due to the effect of alkaline states adopted for scouring. It can possibly cause rise in the hydrophilicity of wool fibres and thus improve its dyeing further.

At a temperature of 85 °C biocarbonization of the pre-scoured wool using a combination of 3 enzymes results in more improvement in its dyeability using acid as well as metal complex dyes. It can be due to the combined influence of the mentioned enzymes (cellulose, pectinase and xylanase) in the separation of natural impurities(vegetable matter) raw wool. Uneven dyeing results from any natural impurities in wool because of its their non-protenic nature. Such natural impurities do not get coloured using acid as well as metal complex dyes and are thus considered as niches that hamper dyeing. More enhancement in the dyeing temperature to 95 °C resulted in restricted increase in the dyeability of the scoured as well as biocarbonized wool using acid as well as metal complex dyes in the proportion of 1:1. When scoured wool has been treated with sulphuric acid it resulted in significant reduction in its dyeability with the investigated anionic dye. It can be because of sulphatation of certain amino acids such as threonine and keratin. Such areas were previously mentioned as dye-resist niches for acid dyes.

8.5 Kinetics of Dyeing

In the case of raw, carbonized, biocarbonized and scoured wool fibres the values of dyeing parameters calculated wool fibres dyed using acid as well metal complex dyes have been determined.

In the case of post scoured wool a decrease in half dyeing time is noticed. It results again because of the increase in the wool wettability followed by elimination of the hydrophobic greasy matters due to scouring. There is almost no effect on the half dyeing time due to the carbonization of the scoured wool by sulphuric acid. The least half dyeing time is achieved with wool biocarbonization using enzyme combinations for all cases.

When particularly dyeing at 85 °C it has been found due to scouring, carbonization and biocarbonization of wool increases its specific dyeing rate constant. In the case of enzyme treated dyed wool there is increase in the diffusion coefficient compared to the untreated wool depending on the nature of the treatment. In the case of scoured wool fibres the 3 enzyme combinations the dyeing rate constant as well as the diffusion coefficient are comparatively higher that that with fibres treated using sulphuric acid.

8.6 Fastness Properties

The fastness to perspiration as well as washing using acid/metal complex dyed for untreated as well as treated wool fibres were measured. There is no appreciable change in the fastness to perspiration and fastness in wool fibres due to scouring, carbonization or bio-carbonization. The findings are almost the same as that achieved on dyeing with Acid Red EG/Neolan Red P at the two specified temperatures.

9 Conclusion

Potato juice has been combined with polyphenolic compound catechol for dyeing in a single step process. The depth of color produced is found to be significantly dependent on temperature of treatment. There is no significant influence on the color obtained by using concentration of the individual components used for coloring. Optimization has been achieved in the wool coloration by means of in situ reaction of diphenolic catechol with enzymes found in potato juice. The formation of color occurs at low temperature and avoids use of auxiliary chemicals. Pinecone which is an agricultural by product is suitable for wool dyeing. It enables to achieve a broad spectrum of colors such as beige, pale to dark green and brown shades having sufficient fastness. Thus because of its reusable and many desirable characteristics pinecone colorant holds great promise. Pinecone is a low cost natural agricultural by product and is found to have great affinity for wool. Dyeing of wool with fermented kum dye exhibits good results in respect to washing and rubbing, tensile strength and elongation percentage and thus proves beneficial to textile industries. It is recommended that the cold traditional kum dyeing technique can be replaced by the heating process following the traditional technique. It is found to give good properties not only for wool but also for silk and cotton yarns which establishes its versatility. The use of a combination of natural and synthetic mordants with leaves of Symplocos racemosa can be effectively used to dye wool to achieve a broad range of soft and light colors. The results achieved with various premordanting agents prove encouraging for adoption in textile industry. Kinetics and thermodynamics of natural dye (A. Nobilis Rech.f.) on wool relates well to partition mechanism confirming that the dye is absorbed by wool as a disperse dye. The dye shows good affinity to wool. The heat of dyeing is found to be negative and the dyeing process appears to be exothermic. The entropy is also found to be negative. Wool cross linked with biopolymer keratin exhibits improved dyeability. Since keratin is a biodegradable protein it can be prospective as a substitute for other unacceptable chemicals used traditionally in wet processing of wool. More investigations are on for application of keratin on certain synthetic fibres to impart certain desirable dyeing properties. Wool has been effectively biocarbonized by purifying from vegetable matters using a mixture of cellulose, pectinase, and xylanase enzymes without affecting its inherent properties, and also does not result in loss in fibre weight and strength. In the case of heavily contaminated wool fibres such enzymes can be combined with mechanical methods or even less aggressive chemical methods. The above discussions clearly show that use of natural dyes and mordants avoid the harmful effects of the chemicals and also enable to effectively utilize the abundantly available natural resources and hold promise of commercialization in the near future.

References

1. Hadzhiyska H, Calafell M, Gilbert J, Daga JM, Tzanov TZ (2006) Biotechnol Lett 28:755
2. Deo HT, Paul R (2003) Int Dyer 188:49
3. Mansour R, Ezzili B, Farouk M (2013) Fibres Polym 14:786
4. Mongkholrattansit R, Keysteufek J, Weiner J (2009) J Nat Fibres 6:319
5. Akimpou G, Rongmei K, Yadava PS (2005) Indian J Tradit Knowl 4(1):33
6. Sharma HM, Devi AR, Sharma BM (2005) Indian J Tradit Knowl 4(1):39
7. Lunalisa P, Ningombam S, Laintonjam WS (2007) Indian J Tradit Knowl 7(1):141
8. Singh NR, Yaiphaba N, David T, Babita RK, Ch DB, Singh NR (2009) Indian J Tradit Knowl 8(1):84
9. Ningombam DS, Ningthoujam SS, Singh PK, Singh OB (2012) Ethnobot Res Appl 10:561
10. Gaur RD (2008) Nat Prod Rad 7(2):154
11. Dayal R, Dhobal PC (2001) Colourage 48:33

12. Martinez MJA, Benito PB (2005) Studies in natural product chemistry: bioactive products (Part K). In: Rehman AU (ed). BV Elsevier, New York, p 303
13. Onifade AA, Al-Sane NA, Al-Zarban SA (1999) Biores Technol 66:1
14. Steiner RJ, Kellems RO, Church DC (1983) J Anim Sci 57:495
15. Weigmann HD, Kamath YK, Ruestch SB, Busch P, Tesmann H (1990) J Soc Cosmet Chem 44:387
16. Kim WM, Kendhal M, Sherman L (1990) US Patent 4, 906, 460 (to Sorenco, Salt Lake City, Utah)
17. Remadnia A, Dheilly RM, Laidoudi B, Queneudec M (2009) Constr Build Mater 23:3118
18. Timmons SF, Blanchard CR, Smith RA (2000) US Patent 6, 159, 495
19. Shikura T, Izumi N, Matsumoto S (1994) Japanese Patent 6
20. Datta MS (1993) J Ind Leather Assoc 43:297
21. Van Dyke ME, Blanchard CR, Timmons SF, Siller-Jackson A, Smith RA (2001) US Patent 6, 270, 791, B1
22. Sehgal PK, Sastry TP, Mahendrakumar M (1986) Leather Sci 33:333
23. Balint B, Bagi Z, Toth A, Rakhely G, Perei K, Kovacs KL (2005) Appl Microbiol Biotechnol 69:404
24. Cardomone JM, Martin JJ (2008) Macromol Symp 272:161
25. Li K, Xu F, Eriksson K (1999) Appl Environ Biol 65:2654
26. Robles A, Lucas R, De Cienfuegos AG, Galvez A (2000) Enzyme Microbiol Technol 26:484
27. Tzanov T, Silva C, Zille A, Oliviera J, Cavao-Paulo A (2003) Appl Biochem Biotechnol 111:1
28. Blanco CD, Gonzalaez MD, Daga Monmany JM, Tzanov T (2009) Enzyme Microbiol Technol 44:380
29. Kim SY, Zille A, Murkovic M, Guebitz G, Cavacopaulo A (2007) Enzyme Microbiol Technol 40:1782
30. Kim S, Lopez C, Guebitz G, Cavacopaulo A (2008) Eng Life Sci 8(3):324
31. Guimaraes C, Kim S, Silva C, Cavacopaulo A (2011) Biotechnol J 6:1272
32. Lokra S, Schuller RB, Egelandsdal B, Engebretsen B, Straetkvern KO (2009) LWT Food Sci Technol 42:906
33. Thygesen PW, Dry IB, Robinson SP (1995) Plant Physiol 109:525
34. Malmstrom B, Ryden L (1968) Biological oxidations. T-Singer Interscience Publications, New York, p 419
35. https://biochemchronicles.wordpress.com/2013/04/14/enzymatic browning
36. Rout Mayer MA, Ralambosoa J, Phippon J (1990) Phytochemistry 29:435
37. Lee CY, Whitaker JR (eds) (1995) ACS sympossium series 600. American Chemistry, Washington, p 49
38. Bradford MM (1976) Anal Biochem 72:248
39. Natarajan SK, Gupta D (2016) Biomimetic coloration of wool using plant juice. Indian J Fibres Text Res 41:306
40. Deo HT, Paul R (2003) Int Dyer 188:49
41. Mansour R, Ezzili B, Farouk M (2013) Fibers Polym 14:786
42. Mongkholrattanasit R, Krystufek J, Weiner J (2009) J Nat Fibres 6:319
43. Ekarmi E, Mafi M, Saberi Motlagh M (2011) World Appl Sci J 13:996
44. Saravanan P, Chandramohan G (2011) Univ J Environ Res Technol 1:268
45. Crews PC (1987) Study Conserv 32:65
46. Kilic A, Hafisoglu H, Tumen I, Donmez IE, Sirvrikaya H, Hemming J (2009) Euro J Wood Product
47. Sakagami H, Kawazoe Y, Komatsu N, Simpson A (1991) Anticancer Res 11:881
48. Momsilovic MZ, Onjia AE, Purenovic MM, Zarubika AR, Randelovic MS (2012) J Serb Chem Soc 77:761
49. Berraksu N, Ayan EM, Mahmoodi NM, Hayati B, Arami M (2011) J Color Sci Technol 5:243
50. Gharachorlou A, Kiadaliri H, Adeli E, Alijaanpoor A (2010) World Appl Sci J 8:334
51. Yaneva ZL, Georgieva NV (2012) Int Rev Chem Eng 4:127
52. Torskngerpoll K, Anderson OM (2005) Food Chem 89:427

53. Saeed S, Meghdad KM, Somayeh S, Sayed MM (2016) Properties of wool dyed with pinecone powder as a by–product colorant. Indian J Fibres Text Res 41:173
54. Adeel S, Ali S, Bhatti IA, Zhisla F (2009) Asian J Chem 21(5):3493
55. Samanta AK, Agarwal P, Konar A, Datta S (2008) Int Dyers 193(4):25
56. Mutua B (1997) Traditional textiles of Manipur. Mutua Museum, Keishampat
57. Mutua B (2000) Tribal Hand woven fabrics of Manipur. Mutua Museum, Keishampat
58. Singh NR, Traditional dyeing skills of the Meities (2003) Proceedings, National seminar on science, philosophy, and culture in Manipur language and literature. Manipur University, Canchipur, Manipur, India, p 14
59. Akimpou G, Rongmei K, Yadava PS (2005) Indian J Tradit Knowl 4(1):33
60. Sharma HM, Devi AR, Sharma BM (2005) Indian J Tradit Knowl 4(1):39
61. Lunalisa P, Ningombam S, Laitonjam WS (2007) Indian J Tradit Knowl 7(1):141
62. Singh NR, Yaiphaba N, David Th, Babita RK, Devi BCh, Singh NR (2009) Indian J Tradit Knowl 8(1):84–88
63. Ningombam DS, Ningthoujam SS, Singh PK, Singh OB (2012) Ethnobot Res Appl 10:561
64. Deb DB (1961) Bull Bot Surv India 3:253
65. Handbook of textile testing, 1st edn. Bureau of Indian standards, New Delhi, 1982, Revised edition 1986
66. Khan MR, Omoloso AD, Kihar M (2003) Fitoterapia 74(5):501
67. Khan MR, Khan M, Srivatsav PK, Mohammed F, Colorage F (2006) 56(1):61
68. Kongkachuichay P, Shitangkoon A, Chingwongomen N (2001) Sci Asia 28(2):161
69. Samanta AK, Agarwal P (2009) Indian J Fibres Text Res 34:384
70. Singh SV, Purohit MC (2012) Univ J Environ Res Technol 2(2):48
71. Crews PC (1982) J Am Inst Conserv 21(2):43
72. Samanta AK, Agarwal P, Datta S (2006) J Inst Eng (India) Text Eng 87:16
73. Chattopadhyay DP (2009) Color Technol 125(5):262
74. Forouharshad M, Montazer M, Yadollah R (2013) J Text Inst 40(3):6
75. Chowdhury SN (1984) Mulberry silk industry. Directorate of sericulture and weaving, Govt. of Assam, Assam, India, p 175
76. Hojo N (2002) Structure of silk yarn biological and physical aspects. Oxford and IBH Publishing Co. Pvt. Ltd., p 96
77. Syed U, Ahmed RJ (2011) Mehran Univ Res J Eng Technol 32(1):133
78. Gogoi A (1999) Indian Text J 107(11):21
79. Gohl EPG, Vilensky LD (1993) Textile for modern living, 5th edn. Longman Cheshire, p 120
80. Gaur RD (2008) Nat Prod Rad 7(2):154
81. Dayal R, Dhobal PC (2001) Colourage 48:33
82. Nishida K, Kabayashi K (1992) Am Dyest Rep 81(9):26
83. Vinod KN, Puttaswamy T, Gowda KNN, Sudhakar R (2010) Indian J Fibre Text Res 35:159
84. Kumaresan M, Palanisamy N, Kumar PE (2011) Int J Chem Res 2(1):11
85. Gulrajani ML, Gupta D (1992) Introduction to natural dyes. Indian Institute of Technology, Delhi, P 81
86. Anderson B (1971) Creative spinning, weaving, and plant dyeing. Angus and Robinson, Singapore, P 24
87. Bains S, Kaur K, Kang S (2005) Colourage 52(5):51
88. Anitha K, Prasad SN (2007) Curr Sci 92(12):1681
89. Martinez MJA, Benito PB (2005) Studies in natural product chemistry: bioactive natural products (Part K). Rehman AU (ed). BV Elsevier, New York, p 303
90. Mahangade RR, Varadarajan PV, Verma JK, Bosco H (2009) Indian J Fibres Text Res 34:279
91. Samanta AK, Aggarwal P (2009) Indian J Fibres Text Res 34:384
92. Vankar PS, Shankar R, Wijayapala S (2009) J Text Appar Technol Manag 6(1):30
93. Dye plants and dyeing (1982) A handbook. Brooklyn Botanical Garden, Brooklyn, p 112255
94. Li K, Wang Z (1995) China Patent CN 1116923 (to Faming Zhuanli Shenquing Gongkai Shoumingshu), 18 March 1995
95. Arora A, Rastogi D, Gulrajani ML, Gupta D Color Technol

96. Arora A, Rastogi D, Gulrajani ML, Gupta D (2012) Indian J Fibres Text Res 37:91
97. Indrayan AK, Yadav V, Kumar R, Tyagi PK (2004) J Indian Chem Soc 81:717
98. Kyu LW, Soo YG (1980) Yakhak Hoechi 24(3–4):151
99. Gupta DB, Gulrajani ML (1994) J Dyers Color 110:112
100. Gupta DB, Gulrajani ML (1993) Indian J Fibres Text Res 18:202
101. Gulrajani ML, Gupta D, Maulik SR (1999) Indian J Fibres Text Res 24:294
102. Bairagi N, Gulrajani ML (2005) Indian J Fibres Text Res 30:196
103. Gupta D, Kumari S, Gulrajani ML (2001) Color Technol 117:328
104. Gupta D, Kumari S, Gulrajani ML (2001) Color Technol 117:328
105. Gulrajani ML, Gupta D, Maulik SR (1999) Indian J Fibres Text Res 24:131
106. Gulrajani ML, Gupta D, Maulik SR (1999) Indian J Fibres Text Res 24:223
107. Das D, Maulik SR, Bhattacharya SC (2008) Indian J Fibres Text Res 33:163
108. Gulrajani ML, Bhaumik S, Oppermann, Hardtmann G (2002) Indian J Fibres Text Res 27:91
109. Johnson A (1989) The theory of coloration of textiles. In: Sumner HH (ed). Society of Dyers and Colorists, England, p 255
110. Triegiene R, Musnickas J (2003) Chemija 14(3):145
111. Onifade AA, Al-Sane NA, Al-Musallam AA, Al-Zarban SA (1999) Biores Technol 66:1
112. Steiner RJ, Kellems RO, Church DC (1983) J Anim Sci 57:495
113. Weigmann HD, Kamath YK, Ruestesh SB, Busch P, Tesmann H (1990) J Soc Cosmet Chem 44:387
114. Kim WM, Kendhal W, Sherman L (1990) US Patent 4, 906, 460 (to Sorenco, Salt lake city, Utah)
115. Remadnia A, Dheilley RM, Laidoudi B, Queneudec M (2009) Constr Build Mater 23:3118
116. Timmons SF, Blanchard CR, Smith RA (2000) US Patent 6, 159, 495 (to Keraplast Technologies Ltd., San Antonio, TX, USA)
117. Shikura T, Izumi N, Matsumoto S (1994) Japenese Patent 06, 40,786 (to Japan kokai Tokyo Koho)
118. Datta MS (1993) J Ind Leather Technol Assoc 43:297
119. Van Dyke ME, Blanchard CR, Timmons SF, Siller-Jackson A, Smith RA (2001) US Patent 6, 270,791, B1 (to Keraplast Technologies Ltd., San Antonio, TX, USA)
120. Sehgal PK, Sastry TP, Mahendrakumar M (1986) Leather Sci 33:333
121. Balint B, Bagi Z, Toth A, Rakhely G, Perei K, Kovacs KL (2005) Appl Microbiol Biotechnol 69:404
122. Cardamone JM, Martin JJ (2008) Macromol Symp 272:161
123. Katoh K, Shibayama M, Tabane T, Yamauchi K (2004) Biomaterials 25:2265
124. Tanabe T, Okitsu N, Yamauchi K (2004) Mater Sci Eng C24:441
125. Nakamura AN, Arimoto M, Takeuchi T, Fujii TM (2002) Biol Pharmocol Bull 25(5):569
126. Kantouch A, Allam O, El-Gabry L, El-Sayeed H (2012) Effect of pretreatment of wool fabric with keratin on its dyeability with acid and reactive dyes. Indian J Fibres Text Res 37:157
127. Mozes TE (1988) Text Prog 17(3):1
128. Knott J (1971) Wool Sci Rev 41:2
129. Zhao W (1986) Text Res J 56(12):741
130. Gouveia IC, Queiroz JA, Fiadeiro JM (2005) Proceedings, 11th international wool research conference. The department of color and polymer chemistry, university of leeds, UK
131. Gouveia IC, Queiroz JA, Fiadeiro JM (2008) AATCC Rev 8(10):38
132. Sedelnik N (1998) Fibres Text East Eur:39
133. Rong ZP, Hua ZJ (1990) Proceedings, 8th international wool conference, vol 3. Wool Research Organization of New Zealand, Christchurch, New Zealand, p 195
134. El-Sayed H, El-Gabry L, Kantouch F (2010) Effect of biocarbonization of coarse wool on its dyeability. Indian J Fibres Text Res 35:330

Chapter 3
Green Chemistry in Textiles

P. Senthil Kumar and E. Gunasundari

Abstract Recent years, the textile industry has become the largest environmentally toxic and polluting industry in all over the world due to their usage of unsustainable and environmentally hazardous chemicals and conventional chemical processing techniques. Green chemistry has been made a great impact on the textile industry to overcome these issues. Green reactions are sustainable, eco-friendly, clean, more efficient, and steady under atmospheric conditions, use of harmless solvents and reduce the auxiliaries, bio-processing, environmentally-safe developed, effective processing, reduction of toxic chemicals, the recovery and as well as the reusability of water, chemical and textile. This chapter is mainly focused on the green chemistry in textile to reduce environmental hazards and health problems associated with chemicals and process in techniques used in textile industry.

Keywords Textile industry · Hazardous chemicals · Sustainable
Eco-friendly and green chemistry

1 Introduction

Currently, our environmental worries are increasing predominantly due to people achieve chemistry in various techniques. In general, biochemical processes contain components including carbon, nitrogen, hydrogen, sulphur, oxygen, iron and calcium. These components are abundantly available in the environment. Industries collect these components from approximately each place and spread them in approaches natural processes certainly not may possible. The component like lead, for instance, obtained widely in sediments very insulated which feature not ever closed it inside organisms. However, in recent times, lead is in all places, mainly

P. Senthil Kumar (✉) · E. Gunasundari
Department of Chemical Engineering, SSN College of Engineering, Chennai 603 110, India
e-mail: senthilchem8582@gmail.com; senthilkumarp@ssn.edu.in

E. Gunasundari
e-mail: gunasundarielumalai2@gmail.com

© Springer Nature Singapore Pte Ltd. 2018
S. S. Muthu (ed.), *Sustainable Innovations in Textile Chemistry and Dyes*, Textile Science and Clothing Technology, https://doi.org/10.1007/978-981-10-8600-7_3

in paints, cars, and computers [1]. Several novel artificial molecules in pesticides, plastics and medicines are very dissimilar against the yields from natural chemistry. The sustainability measure has been established nearly all over the world, especially in industries. The present chapter has been explained clearly about the green chemistry in the textile. Nowadays, a number of sustainable textiles are developed and design by various fashion companies to protect environment from toxic solvent and techniques [2].

2 Green Chemistry

Generally, the green chemistry is the creation, design, and use of chemical products and method either to intentionally minimize or to avoid the formation and usage of toxic materials. It has been also defined as environmentally benign chemical synthesis. The major goals of green chemistry are to minimize toxic effect to human and environment through remodeling, man-made or synthetic methods, harmful molecules, and manufacturing processes [3, 4].

2.1 Principle of Green Chemistry

Green chemistry is mainly focused on to avoid risks during the designing step. The removal of risks from the starting of the chemical method has advantageous to human health and the environment [5, 6]. Generally, green chemistry is clearly explained by twelve principles and are described as following:

- It is enhanced to avoid waste compared to the handling of waste later it can be produced.
- Man-made methods have to be developed for increasing the inclusion of entire constituents used in the process.
- Whenever man-made methods are constructed for producing a product with minute toxicity or without toxicity to protect human and the environment.
- Energy needs have to be identified and reduced for their financial and environmental effects.
- Chemical products must be considered to maintain the efficiency of a process through minimizing toxicity.
- Raw material or feedstock must be regenerate instead of depleting anywhere financially and technically possible.
- Unwanted derivatizations need to be avoided when practicable.
- Catalytic reagents are greater to stoichiometric reagents.
- Chemical products must be formed that could not endure in the environment and collapse into harmless degradation products.

- Analytical methods have to be established to permit in process monitoring, on behalf of real-time, and control preceding to the production of harmful materials.
- The consumption of supporting substances including separation agents, solvents, etc., must be formed needless anywhere probable and safe when utilized.
- Substances utilized for the chemical process should be selected to avoid explosion, release, fire and chemical accidents.

3 Non-ecofriendly Material

Non-ecofriendly materials are including non-biodegradable organic material and toxic substances, which can be unsafe to human and environment [7].

3.1 Non-biodegradable Material

Generally, non-biodegradable materials are not easily destroyed by microorganisms, so they do not have biochemical oxygen demand (BOD) [8]. They have only oxygen demand when they act as a chemical agents. BOD is commonly explained as the quantity of dissolved oxygen required for aerobic biological organisms available in water to collapse organic materials existing in a particular specified water sample over a period of time at a specific temperature. BOD value is usually measured in milligrams of oxygen utilized per liter of a sample in 5 days of incubation time at a particular temperature (20 °C).

3.2 Hazardous Chemicals

Hazardous chemicals are commonly called as a physical hazard which implies that it is confirmed that the specific material can create chronic or acute health effects to human. Health hazard contains carcinogenic or poisonous chemicals can affect skin, eyes or mucous membranes and lungs. Based on the chemical behaviour, hazardous materials are classified as oxidizers, flammable and combustible materials and corrosive or reactive materials, however probably depends on toxicity [9]. Hazardous materials may be mostly categorized into two types including toxic heavy metals and volatile organic compounds (VOCs) [10]. Heavy metals are the category of the metals with atomic number (in the range of 22–34 and 40–52), and the specific gravity of elements including lanthanide and actinide series are 4 to 5 times larger compare to water. With respect to toxicity, the variation of metals is mainly based on the chemical properties metal and their combinations and based on the biological properties of microorganism at hazard. Some of the health hazards associated metalloids and metals are such as cadmium (Cd), lead (Pb), mercury (Hg), chromium (Cr), and arsenic (As). These heavy metals can pass into the human body by air, water, and food otherwise through the absorption via the human skin, which has many forming lipid-soluble organometallic compounds that have a tendency to bioaccumulation within the cells and organs, in this manner damaging their functions. Volatile organic compounds (VOCs) are called as organic chemicals having the high vapour pressure in typical atmospheric condition. Due to the high vapour pressure, the huge number of molecules are evaporated and entered the surrounding air. For instance, the boiling point of formaldehyde is $-19\ °C$, so that will evaporate steadily if it is not taken in a completely closed container. Naturally and synthetically forming chemicals are harmful to human and environment. They may develop various health problems like a headache, eye irritation, nausea, throat and nose irritation, kidney damage, liver damage and central nervous system (VOCs) problem [11].

4 Designing Safer Syntheses

4.1 Green Solvents

Green solvents or bio-solvents are solvents that can be obtained by the processing of agricultural crops. These solvents are substitute to conventional solvents utilized in the chemical processes and will be the new idea to minimize the environmental impact [12]. The advantages of green solvents are given as follows:

- Non-carcinogenic in nature
- Corrosive resistant
- Biodegradable
- Non–ozone depleting
- Excellent solvent properties
- Formed from renewable resources

Examples of green solvents are such as bioethanol, ethyl lactate, polyether, dibasic ester, terpene, organic acid and the siloxane polymer. Bio-solvents are generally derived from renewable resources like the production of ethanol through fermentation of sugar-comprising feedstock, lignocellulosic substances or starchy substances. These are alternative of petrochemical solvents that can minimize the emission of CO_2 into the environment [13–16].

Ethyl lactate is one of the green solvent, which is produced by using corn. It is also called as the ester of lactic acid. Generally, lactate esters solvents are utilized in different industries including coating and paint manufacturing industry due to its advantages such as completely decomposable, ecofriendly, non-ozone depleting, non-corrosive, non-toxic, and recyclable. Ethyl lactate is mostly a solvent in the coatings industry because of its high boiling point, high solvency power, low surface tension and low vapor pressure. It is used as a coating solvent for polystyrene, wood, and heavy metals and furthermore used as a paint stripper and graffiti remover.

Instead of solvents like NMP, xylene, acetone, and toluene, ethyl lactate has been used in the workplace to create a safer environment. For the polyurethane industry, it is used as an effective cleaner, which can able to dissolve an extensive variety of polyurethane resins due to its high solvency power. Ethyl lactate has been used to wash a wide range of metal surfaces, powerfully eliminating solid fuels, adhesives, oils, and greases.

Other supportive research areas in the substitution of the application of VOCs in the industry have the application of ionic liquids and supercritical carbon dioxide used as alternative solvents.

Supercritical fluids ($scCO_2$) are most widely used as a green solvent for several applications. For example, in polymer processing, $scCO_2$ is used instead of CFCs that can minimize ozone depletion [13]. But, scCO2 are also having some disadvantages and are as follows:

- High investment costs;
- Needs to be taken of the safety aspects of the $scCO_2$ equipment; and
- CO_2 losses into the atmosphere during its operation so the process is not completely environmentally harmless.

Super critical carbon dioxide is having properties among that of a liquid and a gas. It can be broadly used as a dry cleaning solvent, which is very easy to remove after a reaction due to its volatility.

Ionic liquids (ILs) are salts in the liquid state that entirely composed of ions. Consequently, melted sodium chloride is generally an ionic liquid although a mixture of sodium chloride present in water is called an ionic solution or a molecular solvent. The common solvents like water, benzene, and ethanol etc., are commonly calling it as molecular liquids whether polar or non-polar. A new group of liquids named room-temperature ionic liquids (RTILs) generally salts which are the liquid over a broad range of temperature and melt below about 100 °C. In general, RTILs is comprised of ions and they act variously from molecular liquids, if they are utilized as solvents. Nitrogen-comprising inorganic anions and organic cations are the major constituents in RTILs. The most generally considered systems have imidazolinium, phosphonium or tricapryl methyl ammonium cations, as well as changing heteroatom functionality [14]. These RTILs are regarded as a green reaction media as they are low-viscosity liquids with certainly not calculable vapour pressure and high thermal stability. They are greatly conductive and have better dissolving power for an extensive range of organic and inorganic compounds. Ionic liquids are used in four major areas such as chemical synthesis, catalysis, electrochemistry, and separation science. Ionic liquids have formed specific scientific interest in phase transfer catalysis and separation or extraction technologies. They can give considerably to green chemistry and the advance of green technology but they contain some disadvantages. The disadvantages of the ILs are explained as follows:

- More toxic compare to other solvents.
- Energetically and economically costly during the preparation these solvent from raw materials.
- Difficult to separate ILs from solutes.

4.2 Catalysis

In general, green chemistry can be divided into three major areas and are as follows:

- Alternative synthesis methods,
- Alternative reaction conditions and
- Alternative chemicals-nontoxic or less risk.

Catalysis leads a major role in the green synthesis, although it's a significant research field in its own right. It can assist not only to form chemical processes greener also to minimize the production cost and environmental effect.

The major objectives of catalyst development come together by way of the principles of green chemistry:

- Make a quick, lifelong, and greatly choosy catalyst that functions in mild conditions. As a catalyst it redeveloped afterward a reaction, single molecule of a catalyst can achieve various conversions.
- To obtain great yields after a reaction, rather, simply a little quantity of catalyst required.

Usually, for carbon-carbon bond-forming reactions, transition metals such as platinum, palladium, and ruthenium are used to prepare catalysts. But, in the meantime, these metals are very costly and available in the small amount in the Earth's crust. Sometimes, to control the selectivity of a reaction, big ligands are essential for the catalysts that can be measured wasteful as said by the green chemistry principles [4]. Hence, investigators try to develop the similar functionality of these catalysts besides an easily obtainable and green metal: iron. This green metal (iron) catalysts can be used an extensive variety of cross-coupling reactions, however, a lot of those reactions need inflammable Grignard reagents. Another side, the iron catalyst may form an alkylsilane which can be applied in shampoo and to become softer denim along with better selectivity and activity compared to platinum catalysts.

A new specific research area is nanoscience that can deliver visions to develop green metal catalysts also. Acid catalysts with silica can minimize the liquid waste made in neutralizing and quenching a reaction. Gold-palladium nanocatalysts may produce hydrogen peroxide, a supportable oxidant to the green chemistry. Catalysts enclosed to magnetic nanoparticles are simple to create, segregated and reprocess the catalyst later a reaction. Some kinds of catalysts may decrease hazards. Phase transfer catalysts (frequently ammonium salts) bring an insoluble reactant among the aqueous and organic phases in a mixture. Tetra butyl ammonium bromide can be used to catalyze the shift of a chloride through a cyanide ion. Heat released during the reaction, however, scientists may try to control exotherm and avoid a temperature spike using governing the stirring speed. Enzymes are generally utilized as catalysts in various industries, predominantly in the pharmaceutical industry; subsequently, they perform at atmospheric temperatures and pressures in water. The three-step enzymatic way to the significant chiral building block employed to create the effective constituent for the cholesterol-lowering drug Lipitor. More than 90%

isolated products, the desired intermediate is produced by the enzymatic process. In whole process, an E-factor is 5 times lesser than a representative pharmaceutical constructed with the green chemistry principles. However, compare to chemical catalysts, biocatalysts are not basically "greener". Enzymes are also used in the paper industries. It can be used to produce a stronger finished product and also let to improve production.

4.3   Green Chemicals

The ecosphere is a closed system using the minimal resources of energy and raw materials and poor capability to accumulate or assimilate the contaminants. Hence, abandoned use of the water, air, and resources may lead to irreversible degradation and even global catastrophe. The improvement of environmentally-enhanced ways and the creation of green chemicals are two sides of green chemistry.

Green chemicals must endure the following measures:

- Produced from easily presented and sustainable resources using ecofriendly processes.
- Minimal tendency to sustain fast, powerful, changeable reactions including explosions which may affect the environment and human.
- Lesser toxicity and non-toxic.
- Non-flammable or poorly flammable.
- Biodegradable.
- Minimal tendency to endure bioaccumulation in food chains in the surroundings.

Dichlorodifluoromethane is the minimum toxic synthetic compounds and is not green since it is very steady and remaining in the atmosphere. It may be source for stratospheric ozone destruction. Hydrofluorocarbons and hydrochlorofluorocarbons are greener substitutes, which is not last long if discharged into the atmosphere or is not have ozone-damaging chlorine. Stable bonds provide persistence and ultimate cause environmental harm. Sodium stearate is a green compound that can be produced by reacting by-product animal fat using sodium hydroxide, which is made through passing an electrical current via salt water. Sodium stearate and calcium in water are reacted to produce calcium stearate. This calcium stearate is non-toxic and biodegradation.

4.4   Greener Energy

Energy conservation is the best approach to achieve a green environment by consuming a lesser amount of resources. The processes including heating, cooling, stirring, distillation, separation, compression, pumping etc., need energy in the form of electrical energy that can be achieved by fossil fuel burning. During this process, carbon dioxide discharged into atmosphere that may cause global warming. Green chemistry has been important in emerging the alternatives for energy generation (photovoltaics, bio-based fuels, hydrogen, fuel cells, etc.) besides to keep the route on the way to energy effectiveness using catalysis and product layout at the forefront. Carbon dioxide is a greenhouse gas (GHG) which is a major concern due to human activities leading to a gradual rise in the atmospheric carbon dioxide level [17]. This recommends that may ultimately change the global climate. Fossil fuel burning is the main supplier to the global emissions. From the green chemistry approach, the combustion of renewable fuels is generally highly needed compare to the combustion of fossil fuels. The biodiesel is the renewable fuel, which is formed from plants oil, for example, soya-beans. It is produced from fats imbedded in plant oils through eliminating the glycerin element that can be a useful resource for the soap manufacturing. The burning of biodiesel will not produce sulfur elements and commonly will not raise the quantity of carbon dioxide present in the atmosphere. Biomass is biological substance from dead or freshly active organisms, like wood, waste, alcohol fuels, and forest residues. It is commonly the plant material grownup to generate electricity or heat. As much as promising fossil derived feedstock should be replaced and regenerating feed stocks. The biomass is also used for production of petroleum, solvents and chemical compounds.

4.4.1 The Advantages of Green Chemistry

The advantages of approving green chemistry are as follows:

- Improved incomes by conserving solvents, energy, reagents, waste, increasing production, disposal expenses, and human resource expenses.
- General application of any industrial process contains basics specifically solvents, raw materials, separation/decontaminations, and chemical reactions using the green chemistry.
- Green chemistry frequently stays unchanged for extended periods of time.
- Innovative separation methods like carbon dioxide extraction, evaporation, phase separation, reforming by-products into new products minimize waste production, and membrane separation. But, an absolutely green process is not be truly green unless used in correct conditions [16, 18].

4.4.2 Challenges in Green Chemistry

The challenges in the green chemistry have been explained as given as:

- The recognition of recyclable feedstock, rather non-food plants and its complete transformation to valuable yields.
- Reactions containing lesser ecological impact as the application of sustainable organic catalysts.

- Manufacturing reactors and processes possessing highest efficacy and lowest waste generation.
- Yields of lowered harmfulness and better biodegradability.
- Phasing out of the flammable, toxic and volatile solvents contaminating atmosphere and estimation of cleaner solvents as substitutes.

5   Textile Production and Contamination

In worldwide, the textile industry is the one of the harmful industry. The consumption of rayon in the textile industry for clothing that destroy forests rapidly. Petroleum-based synthetic dyes and fiber are non-biodegradable and unsustainable. For cotton production, huge amounts of herbicide, fertilizers and pesticides are required.

5.1   Water Utilization

In the textile industry, generally, the large amount water is consumed in wet processing of cellulose fibers for every year. If the usage of water can minimize in wet processing, then that the recovered water can be used for people and also for some other purposes. In dyeing process, water can be saved by various ways and are explained as follows:

- Recover and reuse dye house water
- Optimize the washing and soaping processing
- Minimize the reprocessing
- Minimize the liquid ratio.

5.2   Pollution Problems in the Textile Wet Processing

Several environmental problems are arising in each step in the textile wet processing and are explained clearly as:

- Chemical based wet processing including bleaching, scouring, mercerizing, printing, dyeing etc.
- Utilization of metals noticed in dyestuffs supplementary and binding.
- Remaining dye caused by low dyes fixation and chemicals present in effluent wastewater.
- Formaldehyde observed in the resins, dispersing agents, colorant fixation, and printing paste.
- Use of polyvinyl chloride (PVC) and phthalates in plastisol printing paste.
- Dye effluent wastewater problem.

Chemicals utilized in textile processing can be separated from effluent wastewater by using membrane technology. The major complicated contaminant in effluent wastewater is the dyestuff. Dyestuffs are generally not decomposed in chemicals, water, and light. But, they can be degraded in water by using some treatments techniques (physical, chemical, and biological approaches). The oxidative process is one of the treatment process, which is used to decompose the dye molecules. In this process, hydrogen peroxide and water are mixed and activated by using ultraviolet light to oxidize the dyestuff. But, in this process, poisonous sludge is produced, which is need to be disposed or destroyed by incineration.

5.3 Green Chemistry in Textile Industry

The life cycles of clothing and textiles become unsustainable by the following five issues and are

- Utilization of water—the textile production is considered by the intensive consumption of water and the broad spectrum of processing chemicals. Thus, textile industrial effluents are described by great chemical oxygen demand (COD) and the existence of non-biodegradable constituents like pigments dyes and recently created sizing polymers. The incidence of heavy metals may also come across in several situations.
- Consumption of non-renewable energy in textile production—non-renewable energy resources are underrated and are consumed non-regulatorily. The biggest environmental impact of textiles happens during their usage by consumers (in the range from 75 to 95% of overall environmental impact) and is mostly accounted for the consumption of electrical energy to boil water for running laundry and to dry materials after laundering.
- Consumption of chemicals—the herbicides and pesticides in agriculture and the toxic chemicals in production are consumed uncontrollably.
- Waste generation—an enormous amount of wastes are generated. Non-renewables are not to be and renewable and recycled, which is needed to be composted as much as possible.
- Energy usage for transport—it can be used to take advantage of inexpensive labor, land etc. the production units are distant from consumer place subsequent use of unnecessary non-renewable fuel in transport.

The success of distributing sustainable textile products is reduced if it is packed with the enormous quantity of plastics and layers of foam. Recyclable and environmentally friendly packaging materials will increase the sustainability of the products. In textile and paper industries, energies are produced to advance greener approaches that effect to the reduction in water, energy consumption, and textile processing time [19].

5.4   Greener Fibers

Cotton has been utilized approximately 38% of the globe's textile consumption. It is extremely exposed to pests, particularly in wet places. Though the production of cotton is controlled to 2.4% of land, 25% of insecticide and 11% of pesticide are used for cultivation. To produce one kg of cotton fiber, 7000–29000 L of water is also required for this 'thirsty' crop [20]. Organic cotton is typically identified as cotton that produced by non-genetically-modified plants. To protect organic integrity, whole post-harvest processing, storage, and transport of organic cotton fiber products must be separated from conventional cotton fiber and must not come in contact with banned materials or other pollutants. The treatments must be adequately achieved with natural dyes, pigments, chemicals, and enzymes. 'Organic linen' is prepared from flax fibers grownup with no the consumption of harmful pesticides and fertilizers. Even though wool is an organic fiber, conventional wool production is usually not identified for its eco-friendliness. In general, the wool has been formed in a natural and green technique.

Lyocell has been formed through regenerating cellulose in an organic solvent (N-methyl morpholine-N-oxide (NMMO) hydrate). This risk-free, eco-friendly NMMO solvent consumed is entirely recycled. The fiber is suggestively highly viable compared to oil-derived synthetic fibers and natural fibers like cotton. For cotton production, large space is required than the eucalyptus trees, from that lyocell is formed. Ionic liquids are used to simplify the process of regenerating the cellulose that acts as the solvent and may be almost completely recycled. A really green ionic liquid can be essential to be sustainable, simple and clean to make, toxic free and biodegradable.

Bamboo fiber has specific and natural functions of anti-bacteria, deodorization, and bacteriostasis. Similar chemical antimicrobial, it is not causing skin allergy. The bamboo fibers are obtained to be softer so they can take dye to deeper shades compare to cotton and having natural antimicrobial properties. Bamboo can grow without pesticides, therefore, is more eco-friendly than cotton and some other fibers. The huge amount of adipic acid (HOOC $(CH_2)_4$COOH) have been consumed for the making of nylon, plasticizers, polyurethanes, and lubricants. In the conventional method, this adipic acid is formed from benzene, which is carcinogenic nature. A green synthesis technique has been developed to produce the solvent like adipic acid by using a less harmful substrate. Moreover, the natural resource of this raw material (i.e.) glucose is almost inexhaustible. This glucose is generally altered to produce adipic acid using an enzyme found from genetically modified bacteria. This green way of production safeguards the human and the ecosystem from dangerous chemical compounds.

Polyurethane polymers are tremendously significant and useful materials having several applications in textile, surface and foams coatings, elastomers and adhesives. The polyurethane polymers are synthesized by avoiding or reducing the necessity for hazardous diisocyanate. A sequence of polyurethanes depends on bis-carbamate diols was synthesized by the *Candida Antarctica* lipase B to catalyze the polyesterification. Various polyurethane polymers have been produced based on diamines for which no diisocyanate occurs. The production of synthetic polymers utilizes the huge amount of petroleum for raw material, and natural and synthetic polymers form the big portion of solid waste. Because of this reason, it is needed to form both polymers and to produce and consume polymers which will biologically degrade while discarding. An enormous food products; range of polymers like cellulose present in cotton and wood, protein in wool and silk, and lignin in wood are available in nature. Except lignin, these polymers are formed and also degradable by organisms like particularly fungi and bacteria.

5.5 Bio-polymers

In recent year, the biopolymers or plastics is, produced by corn, starch sugar, and some regenerating resources. Firstly, soy plastics have been used to make several car parts. Cargill Dow's superb also familiar technology utilize corn for synthesizing polylactic acid. It utilizes up to 50% fewer fossil fuel compared to conventional PLA production processes by petroleum-based feedstocks. It will not yields any harmful wastes, and its final yields are sustainable and biodegradable [21]. Polylactic acid (PLA) is fabricated into fibers, films, and rods due to its great strength which is completely biodegradable and compostable, and then they destroy within 45–60 days. Corn husks from corn plants are cleaned and are cut to produce fibers using suitable processing. The properties and structure of corn fibers are same like the natural fibers such as linen and cotton. Cornhusk fibers properties including greatest softness, adequate strength, high elongation, durability, and high moisture recover can offer distinctive properties to yields formed by corn fibers. The numerous advantages likely to industrial constituents, agriculture, the environment and energy utilizing corn fibers are projected to produce these corn fibers desirable over the presently obtainable natural and synthetic fibers. Natural polymers include the polyhydroxyalkanoate (PHA) esters (alkanoates). They were obtained through fluorescent Pseudomonads and some bacteria that form and save them as stock of carbon and energy. Nowadays, these natural polymers have been produced from genetically engineered plants and are entirely biodegradable.

Polycaprolactone (I) is one of the thermoplastic polymer produced from ring opening polymerization using caprolactone (monomer). This polymer is same as PHAs and completely biodegradable, however, degrades at a lesser ratio than PHAs. It is mostly applied in polymer mixture or as a matrix for decomposable composite material due to its lesser melting temperature about 60 °C.

Electrospinning is the effective technique to the spread of biodegradable fibers, particularly, for the manufacturing of nonwoven biodegradable textiles. The electrospinning depends on organic solvents for the mixture of polymeric materials, but in a meantime, many biopolymers are insoluble in organic solvents. Thus, they are not able to form electrospun by conventional methods. Non-volatile room temperature ionic liquids (RTILs) can give a 'greener' processing substitute via reducing the discharge of toxic volatile compounds into the environment. Polyester fiber is non-biodegradable polymers that create environmental difficulties. Two general forms of recycled polyester specifically:

- Easily melted and re-extruded to produce fibers and
- A multi-stage de-polymerization and re-polymerization to provide superior feature yarn. But, the recycled polyester yarn is not superior compared to virgin polyester and not easy to obtain color stability, mainly in pale colors.

5.6   Recycled Textiles

A large amount of the painfully done textile stocks are discarded, fired or buried releasing ozone-releasing methane gases after usage. When the fibers are not burned completely, air-borne particulates are released because of partial decomposition of the material affecting asthma. The textiles are almost 100% recyclable and are not wasted in textile and apparel industry. They are recycled in the textile recycling industry, which are the old and best recognized recycling industries in the world. The recycled textile material can be used in garments. The classifying groups of textile recycling based on volume are denoted in a pyramid structure. The peak of the pyramid is denoted by 'Diamonds' (1–2%) that contain the great value due to their traditional quality. Polyester fiber is a non-biodegradable polymer that makes environmental problem [22].

5.7   Greener Dyestuffs and Auxiliaries

The ecofriendly or green methods are achieved by:

- Removal of dangerous azo dyestuffs.
- Alternate production of sustainable products.
- Examine ecological resource of natural dyes. Generally, they contain poor to moderate light fastness.
- Simple degradable dyes: Majority of the synthetic dyes are very difficult to degrade and eliminate during effluent treatment. The assimilation of the hydroxy group in the di-(tri) arylmethane dyes in ortho position to the central carbon are so simple and valuable oxidation of the dye via dissolving H_2O_2 at sufficient pH under the great catalytic effect of chloride anion composed with methyltrioxorhenium $MeReO_3$ (MTO) [23].

5.8 Biodegradable Surfactants

Fresh sustainable and biodegradable surfactants have been made by way of reacting dextrins along with fatty acids and their derivatives. They have vastly required physical properties such as good wetting and whitening ability, low foaming, and outstanding biodegradability. Recently, solvosurfactants are resulting from glycerol, a renewable material from biodiesel. They have equally surfactant and solvent properties, which are generally used in degreasing, perfumery, coatings, and inks. Alkylphenolethoxylates (APEs) or APEO have been extensively used for emulsification of hydrophobic liquids otherwise in the dispersion of hydrophobic particles such as colorants, fats or resins present in water. Further interests, nonylphenolethoxylates (NPE) is degraded in biological way to produce NP(EO)1–3, a recalcitrant and enormously fish poisonous metabolite, however biodegradation of green substitutes such as alkyl polyglycosides or fatty alcohol ethoxylates result in the fast and whole biodegradation mechanism important to the end which polyglycolethers and polyglycosides of fatty alcohols and their sulphates, sulfosuccinates or phosphates use a whole base set of green surfactants [24].

5.9 Greener Preparatory Processes

Various greener preparatory processes are explained as follows:

- Cleansing of cellulose using extraction with ionic liquids and carbon dioxide.
- Replacement of chlorine bleaching with non-contaminating oxidant, hydrogen peroxide.
- Great temperature water extraction of lignin.
- Removal of ozone-depleting chemicals like carbon tetrachloride.
- Carbon dioxide-based dry cleaning.

5.10 Photo Bleaching

Using a selective photolysis of the dyed compounds, the cellulosic fabrics have been bleached successfully with many excimer lasers (such as XeF, KrF, and XeCl), a black-light fluorescent lamp, and a low-pressure mercury lamp in the existence of sodium peroxocarbonate ($Na_2CO_3 \cdot 1.5H_2O_2$) or combinations of hydrogen peroxide aqueous solutions and sodium carbonate at room temperature. It is an effective process than the thermal bleaching processes when a XeF excimer laser or a black-light fluorescent lamp is used. Sodium borohydride (NaBH4) provided the greatest bleaching efficiency [25].

5.11   Bio-processing

Bio-processing are mainly based on enzymes and their usage in textile production are explained as follows:

- Enzymatic Desizing is mostly carried out by using amylase bacteria.
- Enzymatic bio scouring is performed by lipase/cellulase enzyme. In this operation, only 30% of water conserved and 60% of energy used, minimum fabric weight and strength loss, improved cloth property and dye brightness.
- Enzymatic bleaching is the process used for the elimination of H_2O_2, conserves energy, water, reduce process cycle, improve the biodegradability, and stable bleaching effect, and avoids toxic chemicals.
- Biopolishing and enzymatic based softeners (Cellulose) etc., produces a gentler hand-feel, dirt free surface, improves shine and minimizes pilling.
- Bio-Stone Washing is carried out with a special cellulose enzyme as an alternative to pumic stones. Cellulose performs by the way of releasing the indigo dye on the surface of denim and this process is called as 'bio-stone washing'. A minor quantity of enzyme has been enough to exchange numerous quantity of pumice stones. The minimum amount of pumice stones consumption are not cause as much of damage to the garment and machine.
- Decolorization of dye effluent from textile processing is carried out using enzyme. In general, Laccase enzymes made from fungi such as *Trametes Modesta*. Fungi are most commonly used for dye decolourization in effluent treatment that will be the main factor for environmental concern [26].

5.12   Greener Dyeing Processes

Greener dyeing processes are obtained by the following enhancement and are as given as:

- Optimize dyeing processes to minimize water, energy, electrical power, steam consumption and process time.
- Optimize dye/chemical expenses.
- Remove reprocessing and shade correction.
- Fast dyeing techniques for polyester developing the improved design of dyeing machines and right dyes.
- Reactive dyeing: The dyeing of cationic cotton is more eco-friendly. For this dyeing, alkali and salt are not required and it can be dyed at reasonably low temperatures with reactive dyes. Cationising agent such as 3-chloro-2-hydroxypropyl-trimethylammonium chloride or copolymer of diallyldimethylammoniumchloride and 3-aminoprop-1-ene and copolymer of 4-vinylpyridine quaternised with 1-amino-2-chloroethane is used.
- Sulfur dyeing: The replacement of harmful sodium sulfide along with nontoxic, biodegradable, economical reducing sugars.

- Right-First-Time dyeing: In this dyeing, the inspection stage is eliminated to a major conservation. RFT processing in the dyeing process is achieved when twenty factors are monitored or controlled.
- Various savings are probable by computerization in textile dyeing and printing and are as follows: Process control can be used to save 10–30% of water and energy along with 5–15% saving in dyes and chemicals. 5–10% of dyes, pigments, and chemicals are saved by auto-dispensing. Computer-controlled weighing and stock-taking are worked to save 10–15% of colorants, and chemicals. Dye color measurement and matching are used to save 30–40% of dyes and pigments.

The novel greener coloration technologies are used in the textile industry to improve sustainability. In batch wise dyeing, poly-functional reactive dyes are used to achieve about 90% dye fixation on cellulosic fibers. Chemical-free denim processing can be used in textile production to achieve sustainability. Cold pad–batch dyeing processes reactive dyes that keep a reawakened interest. Cool trans cold transfer printing process: in this process, reactive dyes are moved from printed paper and fixed at room temperature by the cold batching method on pretreated viscose, cotton, silk, and linen. Water energy are consumed in less quantity in this printing process. Laser technology has been an interesting technique in the textile industry, which can be involved to destroy by fire the dyed denim fabric surface or a couple of jeans on a mannequin to reproduce an original worn appearance. It is very fast (i.e.) less than 15 min required to process a pair of jeans [27].

5.13 Supercritical Carbon Dioxide ($scCO_2$) Dyeing

A supercritical fluid has properties of gas and liquid and comprises of a constituent beyond its supercritical pressure and temperature. The supercritical fluids can accept properties intermediate in the middle of a gas and a liquid rising to fill up its vessel such a gas however accompanied by the density of such a liquid. Carbon dioxide ($scCO_2$) has been most frequently used as a supercritical fluid due to its fire resistance, harmless and low cost. As a result of its sustainable as well as risk-free property, it is the greatest supercritical solvent used in textile dyeing process. The CO_2 is a remaining yield of fermentation, combustion, and ammonia production, therefore, for dyeing, CO_2 is not particularly to be manufactured. The remaining dyestuff and the CO_2 are simply segregated through depressurization and are easy to reuse. During this process, wastes are not generated and the energy-intensive drying is not essential after dyeing. For instance, $scCO_2$ is a non-polar solvent, the dispersing agent is not required during polyester dyeing. As the process works at 120 °C, high pressure equipment require that results in high investment expenses. The supercritical dyeing process was examined experimentally for both reactive and non-reactive dyeing. An Excellent dye fixation on cotton dyed using supercritical carbon dioxide can be reached by adding of mono fluorotriazine reactive dyes and less amounts of acids. A washing step of the cotton after dyeing is not essential to eliminate unfixed dye. But, water-soluble dyes are inadequately soluble in scCO$_2$. The dyeing is, so limited to synthetic fibers using scCO$_2$ soluble disperse dyes. A two-step process was proposed for cleaning of old silk textiles. The fibers and the textile structure were not physically damaged [28].

5.14 Digital Ink-jet Printing

Digital ink-jet printing is one of the sustainable technology. It consumes minimum water and generates minimum waste compare to the conventional process in the textile industry. It has some benefits such as small run printing, prototyping, experimentation, and customization. Without contact, ink is sprayed directly through nozzles on textile material in digital ink-jet printing. This technique is also called as non-contact technology. Different color designs are printed on the surface of textile by design data in a computer file. It is not easy to get the same color what we look at the screen. Then, heat or steam is applied to cure the ink once printing is completed. Various types of dyes used in the digital ink-jet printing are including acid dyes, reactive dyes, disperse dyes, latex ink and pigment ink on the textile materials like polyester, cotton, silk, nylon etc.,

5.15 Greener Finishing Agents

The greener finishing agent is mostly applied as crosslinking agents for DP finishes, N-methylol agents or N-methylol amides in the class of formaldehyde reactants. Formaldehyde is carcinogenic to animals. Various formaldehyde-free DP finishes has been explained as follows:

- Cyclic accumulation of glyoxal together with NN/dimethyl urea.
- Polycarboxylic acids (PCA): the major disadvantage is the loss of tensile strength because of acid-catalyzed cellulose structure cleavage. The main significant PCA are citric acid (CA) and butane tetra carboxylic acid (BTCA). Sodium hypophosphite present BTCA offers the similar level of long-lasting press performance as usual DMDHEU reactant, however it is rather expensive [29].

5.16 Nano Finish

Nanotechnology offers big specific areas that help a high level of functionality and is greatly matching with textiles that also develop high specific surface area. Detoxification of textiles on industrial and domestic scales has a most undesirable ecological impact in the textile life. Nano surface finishes are generally oil and water repellent, clothing and fabric stain resistants, self-cleaning, abrasion resistant, antistatic or antibacterial with the resulting sustained the beneficial lifetime of fabric materials and minimize the necessity of dry cleaning or washing. Some treatments may decrease the requirement for ironing with resulting saving of energy and water. Nanocomposites provide major advantages as flame retardant than conventional flame retardants. Low concentrations of silicate are needed in nanocomposites, consequential in minimum density, less expense, and simplicity of formulation. The materials are more biodegradable as the treatment adds no halogens, phosphates or aromatics. They do not produce improved carbon monoxide and soot during combustion. Dispersion of nanoclay into the polymer matrix suggestively increases properties of a polymer, especially, the flame-retardant, mechanical, and thermal properties. Layered silicates present in bulk polymer create a self-protective wall if exposed to heat. The wall reduces fuel pyrolysis and as well as slows the flame temperature. But, the fibers have high surface area and fuel-rich surface. The polymer nanocomposite can be used in combination with lowered amounts of conventional flame retardants [30].

5.17   Green Composites

Green composites depend on natural (particularly plant) fibers and resins are progressively established for several uses as substitutes to the usual non-biodegradable components resulting from petroleum. Plant based starches, proteins, and fibers are renewable except petroleum. Furthermore, the green composites possibly will be effortlessly composted later their lifetime, finishing nature's carbon cycle. Flax yarn reinforced cross-linked soy flour (CSF) composites are completely biodegradable green composites that have been used in secondary and primary constructions in interior applications [30].

5.18   Plasma Treatments

In plasma treatment, dry treatment is carried out by excited gas phase with negligible consumption of water and low consumption of energy. The treatment may be used for surface cleaning, ablation or etching, grafting, polymerization of the maximum external layer of the substrate. In principle, plasma treatment can be carried on all natural and polymeric fibers for the various uses such as desizing, alteration in wettability, development of affinity and leveling character of dyes, anti-felting finish of wool, wool degreasing and sterilisation of fabrics. Plasma treatment is a waterless, sustainable technique, no waste production during wet-chemical processes. [31].

5.19 Adsorbents

The biosorption is an effective technique for the retaining of cations or organic compounds at minimum concentrations in aqueous solutions compared to the conventional treatment with energetic consumptions. Cationised cotton is sometimes used for the removal of anionic dyes from aqueous solution formed in the textile industry as it is natural, inexpensive and renewable. Even though the sorption capacity of the sawdust is not huge, experimental results offer favorable side for the consumption of sawdust as bio-sorbent in decreasing pollution of textile effluents. Lignin, the third important component of plant biomass (16–33%) after celluloses and hemicelluloses, is an inexpensive natural material, available as a by-product from the pulp and paper industry.

5.20 Coagulant

The chemical coagulation using alum and polyaluminium chloride (PAC) is used for treating wastewaters before the biological treatment. Moreover, the chance of Alzheimer's disease caused by aluminum, the continuing effects of these chemicals on human are unknown. To reduce these problems, biodegradable natural polyelectrolytes are take out from plant or animal life, which is the effective replacements to manmade polyelectrolytes. Naturally forming gum, particularly *Cassia angustifolia* seed is used an effective coagulant aid for both acid and direct dyes. This gum can perform as a functioning alternatives, for synthetic chemical coagulants like PAC [31].

5.21 Air Dye Technology

Air Dye technology governs the usage of dye to textiles with no water consumption. The textiles manufacturing need various dozen tons of water for each pound of clothes. This process utilize air in place of water that support the dyes diffusion. Low energy for this process compares to old-style methods of dyeing. There is no water pollution in the color application. Thus, dangerous waste is not released and water is not wasted.

5.22 Eco-label

Eco-labels have developed as a main tool in marketing to more knowledgeable and 'green' consumers. Labels are voluntary declarations and are a clean attempt to set new 'standard' for label textile products. These aim at developing the market with greater environmental protection fulfilling consumer expectation towards reducing environmental and social impacts. Eco-labels are essential to develop a sustainable and a trustworthy textile industry. Examples of a few eco-labels are Oeko-Tex Standard 100 and Global Organic Textile Standard used globally, Blue Angel, Green Seal, Eco-mark etc., [31].

6 Conclusion and Future Trends

The attention of the green chemistry and clean technology will progressively make possible to increase environmental friendly manufacturing systems. "Just-in-time" production may reduce transport and storage complications in small and intensive factories exchanging the heroes of the 21th century. The future challenge is needed to understand dangers such as weak financial environments, climate change, fluctuating buyer activities, resource insufficiency, weak financial conditions, changing consumer behavior etc., and get appropriate actions to protect its future, safeguard environment increase the survives of its people all around the world. Observing more advanced, we may believe to get several chemical factories gaining benefit of local sources like catalysts, feedstocks, reagents, etc., also it is relatively achievable, as transportation turn out to be a major issue, which production will be mostly in response to regional requirements rationally than worldwide market prospects. Therefore, green can be the most modern fashion in the planet, as technique is continuously developing and, as a significance, publics, inventers, and industries come to be more eco-conscious.

References

1. Historical Production and Uses of Lead (2011) http://www.ldaint.org/
2. Sustainable Materials (2013) http://www.sustainablematerials.org.uk/resource/textiles.html
3. Anastas P (2011) Twenty years of green chemistry. Chem Eng News 89(26):62–65
4. American Chemical Society (2015) History of green chemistry. http://www.acs.org/content/acs/en/greenchemistry/what-is-green-chemistry/history-of-green-chemistry.html
5. Clark J, Macquarrie D (eds) (2002) Handbook of green chemistry and technology Ltd. Blackwell Science, Oxford
6. Hutchings GJ (2005) Green chemistry has a golden future, Europacat 7. Cardiff University, UK
7. Webster's New Millennium Dictionary of English (2006) Preview Edition(v0.9.7). Lexico Publishing Group, LLC, USA
8. Smith PG, Scott JS (2005) Dictionary of water and waste management. Elsevier, Oxford, UK
9. US Occupational Health and Safety Administration (2012) OSHA, Under US Department of Labor. http://www.osha.gov/
10. Phipps DA (1981) Chemistry and biochemistry of trace metals in biological systems. In: Lepp NW (ed) Effect of heavy metal pollution on plants. Applied Science Publishers, Barking

11. US Environmental Protection Agency (2013) An introduction to indoor air quality, volatile organic compounds. http://www.epa.gov/iaq/voc.html

12. Manahan SE (2001) Fundamentals of environmental chemistry. CRC Press LLC, Boca Raton

13. Nalawade SP, Picchioni F, Janssen LPBM (2006) Supercritical carbon dioxide as a green solvent for processing polymer melts: processing aspects and applications. Prog Polym Sci 31:19–43

14. Scammells P, Scott J, Singer R (2005) Ionic liquids: the neglected issues. Aust J Chem 58:155–169

15. Nagendrappa G (2002) Organic synthesis under solvent-free condition: an environmentally benign procedure—I. Resonance 7: 59–68. http://www.ias.ac.in/resonance./

16. Mohammad A, Inamuddin (eds) (2012) Green solvents i: properties and applications in chemistry. Springer, London. ISBN 978-94-007-1711-4

17. Green Power Defined (2012) Green Power Partnership US EPA. http://www.epa.gov

18. Poliakoff M, Licence P (2007) Sustainable technology: green chemistry. Nature 450(7171):810–812

19. Teli MD (2008) Textile coloration industry in India. Color Technol 124(1):1–13

20. Reddy N, Yang Y (2005) Green Chem 7(7):190–195. https://doi.org/10.1039/B415102J

21. Chodak I, Blackburn RS (2009) Sustainable textiles: life cycle and environmental impact. Woodhead, Oxford, UK, pp 88–112

22. Hawley JM (2006) Textile recycling: a system perspective. In: Yong Y (ed) Recycling in textiles. Woodhead, Cambridge

23. Moozyckine AU, Davies DM (2002) Green S as a prototype for an environmentally-degradable dye: the concept of a 'green dye' in future green chemistry. Green Chem 4:452–458

24. Höfer R, Bigorra J (2007) Green chemistry—a sustainable solution for industrial specialties applications. Green Chem 9:203–212. https://doi.org/10.1039/b606377b

25. Ouchi A, Obata T, Oishi T, Sakai H et al (2004) Reductive total chlorine free photochemical bleaching of cellulosic fabrics, an energy conserving process. Green Chem 6:198–205. https://doi.org/10.1039/b315580c

26. Preša P, Tavčer PF (2009) Low water and energy saving process for cotton pretreatment. Text Res J 79(1):76–88. https://doi.org/10.1177/0040517508092019

27. Thiry MR (2010) AATCC Rev 10(3):32–39

28. Höfer R, Feustel D and Fies M(1997) Derivate natürlicher Öle als Rohstoffe fur Lacke und Druckfarben, Welt der Farben, 11–18

29. Andrews BAK (1990) Non-formaldehyde durable press finishing of cotton with citric acid. Textile Chemand Color 22:63–67

30. Jimenez ABY, Bismarck A (2007) Surface modification of lignocelluloses fibers in atmospheric air pressure plasma. Green Chem 9:1057–1066. https://doi.org/10.1039/B618398K

31. Sanghi R, Bhattacharya B, Dixit A, Singh V (2006) Cassia angustifolia seed gum as an effective natural coagulant for decolourisation of dye solutions. Green Chem 4:252–254

Chapter 4
Bioremediation: Green and Sustainable Technology for Textile Effluent Treatment

Luqman Jameel Rather, Sabiyah Akhter and Qazi Parvaiz Hassan

Abstract In recent decades, textile industrial sectors are getting increasing interest worldwide in global contest due to the diverse and changing world market conditions in terms of price, design, ease of handling, durability, and product safety. The increasing ecological and health concerns related to the use of large amounts of dyes (Synthetic as well as natural) in textile industries to counter the growing demands of people lead to the design, development and establishment of new dyeing strategies; and technologies in addition of reducing the load of effluents in wastewaters. Textile industrial sectors and its associated wastewaters have become an increasing cause of main sources of severe pollution worldwide. The effluents produced from these textile wet processing industries are very diverse in chemical composition, ranging from inorganic finishing agents, surfactants, chlorine compounds, salts, total phosphate to polymers and organic products. Most of the techniques used for removal of dye effluents from wastewaters were physico-chemical methods which are costly and cause an accumulation of concentrated sludge. Hence there is need to develop alternative treatments plans and strategies that are are cost effective and environmentally benign. In this paper authors review the advancements in eco-friendly and sustainable technologies used for minimizing the negative environmental impact of wastewater from textile sectors by biological and combination systems.

Keywords Wastewater · Dyeing · Textile dyes · Sustainability · Ecological

L. J. Rather (✉)
Department of Computer Science and Engineering, University of Kashmir, North Campus, Delina, Baramullah 193103, Jammu and Kashmir, India
e-mail: luqmanjameel123@gmail.com

S. Akhter · Q. P. Hassan
CSIR-Plant Biotechnology Division, Indian Institute of Integrative Medicine, Sanatnagar, Srinagar 190005, Jammu and Kashmir, India

© Springer Nature Singapore Pte Ltd. 2018
S. S. Muthu (ed.), *Sustainable Innovations in Textile Chemistry and Dyes*, Textile Science and Clothing Technology, https://doi.org/10.1007/978-981-10-8600-7_4

1 Introduction

Rapid industrialization and urbanization of modern industries result in the continuous discharge of large amounts of wastes/effluents to the environment. However, colored effluents from textile industrial sectors and associated wastewaters have been increasing at a high rate, contributing the world wide pollution significantly. In general colored substances/effluents and in particular synthetic azo dyes are undesirable, not only because of their color, but also because the secondary byproducts of azo dyes are toxic and/or mutagenic in nature [1–3]. In textile dyeing processes considerable amounts of dyestuff (Natural and synthetic) are discharged with the effluents and are currently representing a major ecological concern. Depending upon the limits of dye concentrations in water bodies (1 ppm in the UK) most of the physico-chemical techniques used would require a reduction of the dye concentration by up to 98% [4]. The chemical structures of dye molecules are designed to resist color fading to UV radiations, chemical attack and quite resistant to microbial attack.

Use of synthetic dyes after second half of 19th century has increased because of their cost-effectiveness and wide range of available shades with much better wash and rubs fastness results [5, 6]. However, adverse growth effects on methanogenic bacterial cultures have been reported with azo dyes and this toxicity corresponds mainly to azo functional group rather than to their secondary byproducts (Primary aromatic amines) [7–9]. A wide variety of bacterial isolates and helminthes are reported to cleave chromophoric groups of azo dyes aerobically as well as anaerobically in the human intestinal microflora [10, 11]. More than 100 L of water are currently consumed during processing and finishing of 1 kg of textile material [12]. Thus, there is a huge demand for new environmental friendly effluent recycling technologies/strategies to reduce wastewater problems. However, biological treatments techniques/strategies are much cheaper and easier to operate and have become the main focus for dye degradation and decolorisation in recent studies. Wastewater form textile industrial sectors can be reused for preparation of dyeing baths through microbial or enzymatic dye degradation processes [13]. Due to high specificity of enzymes decolorisation processes, valuable dyeing additives and fibers remain intact and the main target are chromophoric groups of dye molecules. Both bacterial, fungal cultures and specific enzymes are having high potential and bright future in the treatment processes of colored effluents in the textile industry [14].

2 Modes of Bioremediation

The term "bioremediation" covers a wide range of biological and combination processes that use natural resources to control wastewater pollution problems. Xenobiotics, mainly consists of aromatic rings substituted by electron-withdrawing groups, are of anthropogenic origin which displays high persistence in the ecological ecosystems [15]. To reduce toxicity levels, several remediation techniques have been used

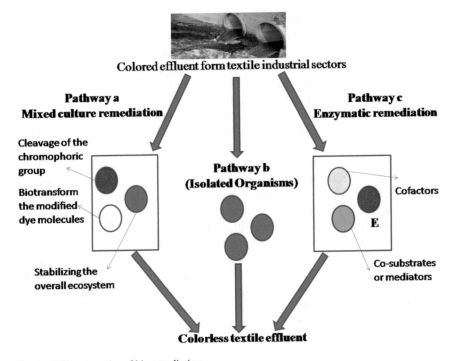

Fig. 1 Different modes of bioremediation

in recent past such as microbial degradation (bacteria and fungi), phytoremediation (plants which involves several biological mechanisms), and enzyme remediation (specific enzymes) to degrade pollutants/colored substances. Biological systems have to be designed in such a way that the dye under such systems should preferably be complexly degraded along with decolorisation. Due to these requirements, there is currently no simple solution but combination of bacterial and enzymatic cultures (mixed cultures) can be effectively used for this purpose. Modification of chromophoric group of dye molecules may in fact be another way of bioremediation through use of single microorganism. However, such decolorisation processes yields metabolic end products that have toxic consequences, such as anaerobic reduction of azo dyes [16]. Hence, the complete degradation of a xenobiotics leading to the release of carbon dioxide, ammonia, and water may turn out the only way for sustainable bioremediation process to be achieved within mixed cultures. In general, mixed cultures usually exhibit higher stabilities towards environmental stress caused by changes in temperature, pH or composition. A classical representation of different modes of bioremediations is presented in Fig. 1.

With mixed cultures (**Pathway a**), there are more than one species involved in remediation process with separate motive with one species (cleavage of the chromophoric group); another (biotransform the modified dye molecules); whereas others helps in stabilizing the overall ecosystem. The biotransformed dye molecule will be more accessible to another organism that otherwise is not able to attack this dye [17, 18]. In this way, the decolorisation could be achieved mutually depending on the presence of several species and on their synergistic action. However, isolated organisms are directly involved in dye biotransformation (**Pathway b**). Alternatively, the third remediation process involves enzyme (**Pathway c**). Their action may get enhanced, depending on the presence of cofactors, co-substrates or mediators.

Depending upon the type of reactor, microbial cell cultures can be fixed in different reactors by means of immobilization [19]. Different bacterial cell immobilization techniques have been used for dye degradation like immobilization through use of mineral material, seashells, or nylon [20]. Calcium alginate and granular sludge have provided support for mixed cultures [21–23]. Whereas, bacterial cells-activated carbon immobilization technique allows simultaneous oxidation of biodegradable contaminants and adsorption of non-biodegradable matter regenerating activated carbon in one single step [24]. Another process for synthetic dye biodegradation involves flocculation of *Pseudomonas* sp. by using aluminum sulfate [25].

Biochemical transformation of the target dye molecule by use of singe cell cultures or enzymes may provide a better bioremediation route as compared to the physical retention of dye on biomass by means of adsorption/ion exchange on the outer surface area of bacterial cell, as contaminated biomass has to be treated separately in next step. A number of single bacterial cultures or microorganisms have been found to decolorize textile dyes including bacteria, fungi, and yeasts [26, 27]. Resistance against toxic effects of dye is another important requirement for an organism to be potent bioremediation agent. Isolated enzyme systems may be preferred in cases where target dye molecule inhibits growth, especially at high concentrations of dyestuff. So, the major problems have to be dealt very curiously as dye house effluents are complex mixtures of salts, detergents, dispergents, metals, weakly adsorbed dyes, and may strongly vary in chemical composition depending on the production charge. Table 1 represents various bioremediation techniques which are helpful for easy degradation and chemical transformation of dye molecules.

Dyes display wide structural variety and thus do possess very different chemical and physical characteristics. They are designed to be very stable to UV radiations and thus extensive reviews related to their degradation have been reported [28–30]. Combinations of chemical and microbial/enzymatic treatment have also been described in recent past [31–35]. In this chapter we review the advancements in eco-friendly and sustainable bioremediation technologies used for minimizing the negative environmental impact of wastewater from textile industrial sectors by biological and combination systems.

Table 1 List of environmentally benign and sustainable bioremediation techniques

S. No.	Bioremediation technique	Species involved in bioremediation	Substrate involved	References
1	Bacterial remediation	Various bacterial species	Aromatic compounds, Azo and Carboxylated azo compound	[19–25]
2	Fungal/algal remediation	Micro-algae, White-rot fungi (Excreted Enzymes), alginate-immobilized algae	Reactive Orange 96, Reactive Violet 5, Reactive Black 5	[19]
3	Enzyme remediation	Extracellular and intracellular enzymes	Anthroquinone, indigoid-based dyes, phenolic azo dyes, Phenolic and non-phenolic aromatic compounds, H_2O_2, reactive azo dyes	[13, 14, 19]
4	Biosorption	Fungal mycelia	Wastewater from textile industries	[8]
5	Phytoremediation	Living green plants, endophyte-assisted poplar tree	Heavy metals, Trichloroethylene	[23]
6	Lichen remediation	Lichen *Dermatocarpon vellereceum*	Navy Blue HE22 (NBHE22), dye mixtures and real textile effluent	[11, 58, 59]

3 Bioremediation Processes

3.1 *Biological*

Biological remediation (Bioremediation) have emerged a new ecofriendly and sustainable technique for treating effluents form textile dye industries at various stages of dyeing and finishing processes. Single cell and mixed cultures of microbial and enzymatic decolorisation/degradation of azo dyes are very significant to address this particular problem because of their environmentally-friendly, inexpensive nature, and non-production of sludge. A wide variety of bacterial cultures including *Proteus* sp., *Enterococcus* sp., *S. faecalis*, *B. subtilis*, *B. cereus*, *Pseudomonas* spp. and even helminthes have been used for azo dye reduction [36, 37]. Azo reduction may be enhanced through addition of certain co-substrates which act as catalysts and serve as reductive equivalents. Depending on the type of organism/microbial culture used for a particular bioremediation process, different additional nutrients may thus be supplied as co-substrates. A wide variety and economical strategies for designing new or improved catalysts for bioremediation have been developed over recent past.

3.1.1 Bacteria Bioremediation

Detoxification and degradation of environmental contaminants/effluents using microbes/bacterial cultures has received increasing attention since recent times to clean up the polluted environment/ecosystems [38, 39]. Aromatic compounds and their secondary metabolites are highly liable to biological degradation under both aerobic as well as anaerobic conditions [40]. The oxidation of aromatic ring system by molecular oxygen from atmosphere can be easily done by the enzymes mono- and di-oxygenase [41]. Azo dyes substituted with nitro and sulfonic groups are quite reluctant to aerobic degradation and are resistant to oxygenases attack probably due to the electron-withdrawing nature of the azo bond [42–44]. However, in the presence of azo reductases (oxygen-catalyzed enzymes), some aerobic bacteria are able to reduce/degrade azo compounds and produce aromatic amines which will be separately discussed in latter part of this chapter. Recently described aerobic sequencing batch reactors are being continuously used and have gained much scientific attention for dye degradation processes [45, 46]. Degradation in presence of ethanol stimulates the respiration of facultative aerobic microorganisms during exposition of methanogenic granular sludge. Consequently, azo dyes have been reduced by methanogenic colonies in anaerobic conditions and thus help in full azo-dye mineralization in aerated anaerobic/aerobic reactors [40, 47].

Flavin-free aerobic azo dye reductases from *Pseudomonas* species strains K22 and KF46 with NADP(H) and NAD(H) as cofactors carried out oxidative cleavage of carboxylated substrates of bacteria and also their sulfonated structural analogues [48]. Another cloned aerobic azo reductase from *P. kullae* K24 used carboxylated azo compound 1-(40-carboxyphenylazo)-4-naphtol for their growth and conserved a putative NADPH-binding site [49]. Oxidative attack mediated by peroxidases with soil bacteria has also been reported in literature [50]. Triphenylmethane dyes have been degraded by *B. subtilis*, *P. pseudomallei*, *Corynebacterium*, *Mycobacterium*, and *Rhodococcus* species giving an indication that dye degradation is solely possible by using aerobic cultures. Moreover, an azo dye degrading extracellular peroxidase is also released by a *Flavobacterium* sp. [51].

3.1.2 Fungal and Algal Bioremediation

Formation of exoenzymes such as peroxidases in presence of H_2O_2 and phenoloxidases in fungi are responsible for azo dye degradation [52]. Lignin and manganese peroxidases have been found to exhibit similar reaction mechanism during their catalytic cycle that starts with the enzyme oxidation initiated by H_2O_2. Lignin peroxidase oxidizes both phenolic and non-phenolic aromatics where as manganese peroxidase oxidizes phenolic compounds by getting oxidized from Mn^{2+} to Mn^{3+}

[53]. Heinfling et al. (1997) tested eighteen fungal strains capable of degrading ligno-cellulosic material or lignin derivatives, against Reactive Orange 96, Reactive Violet 5 and Reactive Black 5 out of which the strains of *B. adusta*, *T. versicolor* and *P. chrysosporium* were able to decolorize all azo dyes [54].

P. chrysosporium and the non-ligninolytic fungi (*Basidiomycete*) have been reported to decolorize various types of dyes and are effective in reducing wastewater problems to a greater extent. However, it has been found that wood rotting fungi such as *A. fumigatus* G-26 and *A. oryzae* are more efficient in decolorizing different types of dyes compared to *P. chrysosporium* [55]. *Aspergillus* strains are applied on the production of cromic acid, galic acid, citric acid, enzymes, isomerases, pectines, lipases, and glucanases. Thus, it is more favorable to add the biomass in biosorption processes for removal of dyes from textile effluents [56]. It may be a profit generating concept to use fungi biomass as biosorbents for the reduction of cost/disposal problems of biomass. From this point of view *Aspergillus* sp. are suggested as an alternative option. Additionally, normal and autoclaved hyphae of *A. oryzae* were used for the removal of Procion Violet H3R and Procion Red HE7B from aqueous solutions. The mechanisms of biosorption are different for structurally different types of dyes and mainly dependent upon their structure and substitutions. Functionalized microbial surface possesses provide better and increased interaction for the removal of hazardous materials from industrial effluents [57].

Micro algae are known to remove dyes from textile effluents by bioadsorption, biodegradation and bioconversion. Microalgae help to reduce eutrophication (degrading dyes) from aquatic ecosystems by removing nitrogen, phosphorus, and carbon content from water [58, 59]. Alternatively, living and non-viable micro algae grows at rapid pace and have been used for the removal of color from wastewaters. Khalaf et al. (2008) reported the removal of reactive dye (Synazol) from textile wastewater by non-viable biomass of *Spirogyra* and living biomass [60]. Living biomass of macroalgae such as *C. lentillifera* and *C. scalpelliformis* are found to be helpful in the removal of basic dyes by biosorption mechanism [61, 62]. *C. vulgaris* removed 63–69% of azo dye tectilon yellow 2G by converting it to aniline [63]. Recently it has been confirmed that removal of color by algae (*Cyanobacteria Synechocystis* and *Phormidium*) can be enhanced by adding growth regulator hormones (Triacontanol) [64].

3.1.3 Enzymatic Bioremediation

Textile dyes (synthetic and natural) represent a massive and diverse variety of chemical compounds. Due to the diversity in structure and function textile dyes have very different dyeing properties which in turn reflect the solubility and reactivity towards the fabric; they are generally degraded/decolorized by only few enzymatic processes. Wide range of substrate specificities and redox-active nature makes enzymes more valuable and environmentally friendly alternatives for dye decolorisation processes. Chivukula and Renganathan (1995) reported the oxidation of electron rich azo dyes

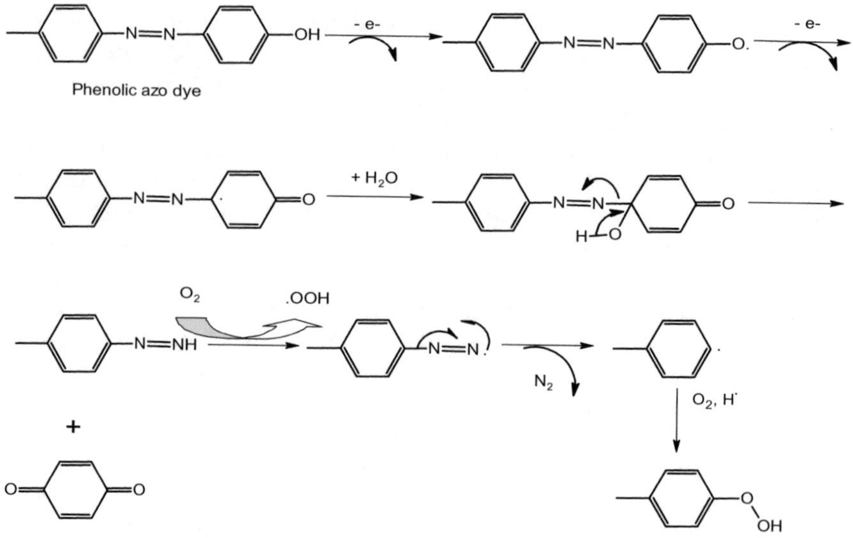

Fig. 2 Mechanistic details for release of molecular nitrogen from phenolic azo dye initiated by laccase of *Pyricularia oryzae*

by laccase of *P. oryzae*. In this enzyme oxidation process molecular nitrogen is released by azo bond cleavage (Fig. 2) [65].

Although, so many reviews have been published regarding the mechanistic and applied aspects of oxidative enzymes in the degradation of xenobiotics [66, 67]. Table 2 summarizes the use of different types of enzymes applied for dye degradation. Phenoloxidases are oxidoreductase enzymes that catalyses the oxidation of phenolic and other aromatic compounds without the use of cofactors [52]. Laccases (copper-containing enzymes) have very broad substrate specificity with respect to dyes [13]. However, it has been found that laccases from *T. versicolor*, *P. pinisitus* and *M. thermophila* decolourise anthraquinone and indigoid-based dyes at higher rates compared to Direct Red 29 (Congo Red) [43]. The redox mediators used have been found to extend their substrate specificity with regard to various classes of dyes [43, 68, 69].

Intracellular mono-oxygenases and dioxygenases are largely present in living systems. They cause degradation via incorporation of oxygen atoms (biohydroxylation), releasing carboxylic acids through ring cleavage mechanisms [70–72]. Laccases have an advantage of using molecular oxygen as co-substrate over oxidoreductases or cytochrome P450 reductases [73]. Similarly, peroxidases only depend on the availability of hydrogen peroxide as co-substrate. Hence, they are promising candidates for bioremediation purposes. In contrast to laccasses and peroxidases, the application of oxidoreductases requiring cofactors like NADH, NADPH, or FADH which are extremely expensive compounds and are not economically feasible. However, decolorisation processes with such enzymes usually take place in whole cell cultures.

Table 2 List of enzymes mostly employed for dye degradation and biotransformation

S. No.	Various types of enzymes involved in bioremediation	Substrate used	References
1	Laccasses, immobilized laccasses	Anthroquinone and indigoid-based dyes, Phenolic azo dyes	[13, 43, 65, 68]
2	Tyrosinases	Phenolic and other aromatic compounds without use of cofactors	[70, 73]
3	Peroxidases, Lignin peroxidases	Azo and anthroquinone dyes, Phenolic and non-phenolic aromatic compounds	[52, 65, 68]
4	Immobilized catalases	H_2O_2 removal	[52, 72]
5	Glucose oxidase	Reactive azo dyes	[73]

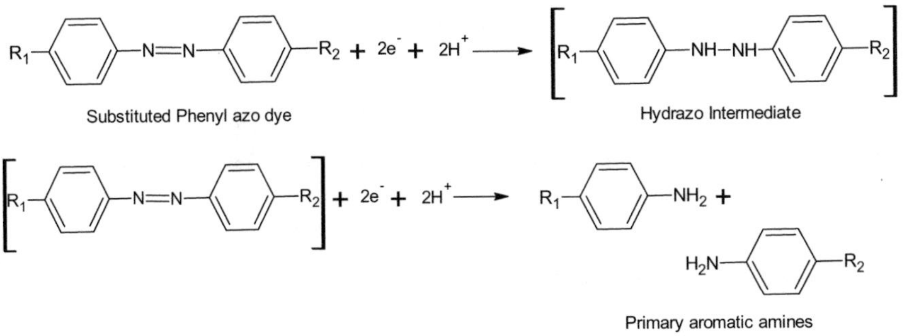

Fig. 3 Schematic representation of anaerobic azo dye reduction mechanism

3.1.4 Bioremediation by Anaerobes/Facultative Organisms

Bioremediation by anaerobic/facultative microorganisms needs low reduction potential (≤ 50 mV) for the effective decolorisation of dyes [74, 75]. Biochemistry of dye decolorisation and reduction is mainly covered by reduction under anaerobic conditions. The cleavage of azo bond (–N = N–) requires transfer of four-electrons, which proceeds under two stages at the azo linkage as shown below in Fig. 3.

However, exact mechanism of azo dye reduction is still obscure whether it occurs intracellularly or extracellularly. Flavin based reducing agents can act as an electron shuttle from NADP(H)-dependent flavoproteins to azo dye [76]. Sulphonated azo dyes are not reduced intracellularly because of limited membrane permeability [77]. However, it was found that addition of toluene increased color removal of sulfonated azo dyes by cell free-extracts as it is a membrane-active compound which increases cell lysis [73]. In extracellular dye reduction mechanism, reduced cytoplasmic cofactors do not contribute to the chemical dye reduction [78]. However, results from cell fractionation experiments carried out by Kudlich et al. (1997) confirms that quinone reductase located in the cell membranes increased reductive decolorisation of sul-

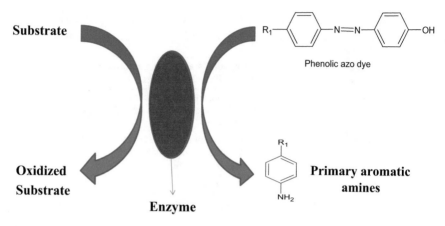

Substrate

R₁ [Phenolic azo dye structure with R₁—benzene—N=N—benzene—OH]

Phenolic azo dye

Oxidized
Substrate

Enzyme

R₁ [benzene ring with NH₂]

Primary aromatic
amines

Fig. 4 Schematic representation of direct enzymatic reduction of azo dyes

fonated azo compound [73]. Cytosolic fraction of *Escherichia coli* showed capacity for azo dye reduction due to presence of NADH-dependent lawsone reductase [79].

3.2 Combination Methods

3.2.1 Biological and Chemical Reductive Decolorisation

The anaerobic reductive decolorisation of azo dyes involves biochemical (biological and chemical) mechanistic pathways [80]. The biological pathways include enzymes called azo reductases, which grow by using azo dye as a source of carbon and energy. However, there is no evidence of anaerobic azo reductases those catalyze the reduction of various substrates including azo dyes [77]. However, the reduction proceed through a co-metabolic reaction (Fig. 4), which involves NAD(H), NADP(H), $FMNH_2$ and $FADH_2$ as secondary electron donor for cleavage of azo bond [76].

The anaerobic chemical reduction involves biogenic reductants like sulphide, cysteine, ascorbate or Fe^{2+} [81]. However, it has been observed that sulfate concentration did not have an adverse effect and donot obstruct electron transfer to the azo dye even up to 60 mM concentration [82]. In another investigation no significant improvement was achieved on the color removal by an aerobic-anaerobic sequencing with sulfate through sulphide reduction (0.35 mM), while testing Acid Orange 7 even though a sulfate reducing microbial population was established and was concluded that the color removal is mainly done through biological process [83]. Low rate of decolorisation (10–30%) under aerobic conditions are fully justified as oxygen is more effective electron acceptor compared to azo functional group, but the actual rates of decolorisation are extremely dependent on the type of dye and the type of substitutions.

However, it has been concluded through experimental results that anthraquinone and phthalocyanine dyes are rather recalcitrant [84–86]. Anaerobic microorganisms are more susceptible to some of the dyes that may lead to permanent loss of the methanogenic activity even at low dye concentrations [86]. This is confirmed by the study of Dos Santos et al. (2005) exerting 50% reduction in methanogenic activity (IC-value was 55 mg/l at 30 °C) for Reactive Black 19 azo dye [84].

3.3 Reductive Decolorisation of Azo Dyes in the Presence of Redox Mediators

Redox mediators/cofactors are the chemical entities that accelerate the rate of electron transfer reactions to several orders of magnitude [84]. Redox mediators are very much effective for reductive transformation of iron, nitroaromatics, polyhalogenated compounds, and radionuclides in addition to usual reductive decolorisation [87, 88]. Keck et al. (2002) showed that the quinoid redox mediators are produced during the aerobic degradation of naphthalene-2-sulfonate (2-NS) by *S. xenophaga* strain BN6, which helps in the anaerobic reduction of azo dye [89]. Similarly FAD, FMN, riboflavin, AQS, AQDS and lawsone, have been extensively used as redox mediators during azo dye reduction [84]. As represented in Fig. 5 below, reductive decolorisation in the presence of redox mediators/co-factors can be represented in two steps, the first step being a non-specific enzymatic reduction, and the second step being a chemical re-oxidation of the mediator by the azo dye. The reduction occurs on account of redox potentials between the half reactions of azo dye and the primary electron donor [82]. However, due to unknown standard redox potential of most azo dyes, polarography may be used to get this information.

The standard redox potential ($E°$) of a compound is the measure of its ability to act as a redox mediator. However, different decolorisation rates have been reported in the presence of mediators with similar E^0 values and same decolorisation rates for redox mediators with different E^0 values [90]. Walker and Ryan (1971) concluded that the decolorisation rates are related to the electron density of azo bond and rates of decolorisation increases by lowering the electron density in the azo linkage (electron-withdrawing groups as substituents) [91]. Hence, use of redox mediators/co-factors would reduce steric hindrance along with accelerating transfer of reducing equivalents to the azo dye [92] and consequently decreases the activation energy of the chemical reaction [84]. Thus, theoretical estimation of decolorisation rates by using specific redox mediators, differences in electro-chemical factors between mediator and azo dye should also be considered.

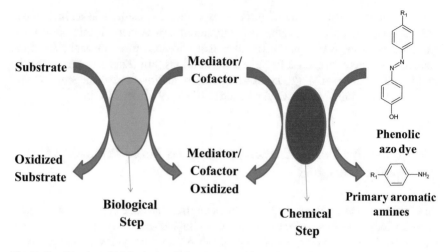

Fig. 5 Combination of biological and chemical steps for indirect azo dye reduction

4 Gene Profiling of Dye Degrading/Transforming Organisms

In the course of natural adaptation, organisms degrading/transforming xenobiotics evolve naturally or under controlled laboratory conditions [93, 94]. Thus, well-directed optimized hybrid strains which can degrade or transform one particular type of dye may be obtained directly via genetic profiling/engineering. However, using cloning procedure and transferring genes for dye degrading enzymes can be designed and possibility to combine the abilities of mixed cultures within one single species can be easily achieved [95]. Successful decolorisation of an azo dye was achieved using *E. coli* carrying the azoreductase genes from a wild-type *P. luteola* [96]. This approach could be useful alternative for time management to adapt appropriate cultures and isolated strains. Alternatively, heavy-metal resistance genes can be implanted into the dye-degrading organisms and can be used to reduce environmental toxicity of such ions in textile bioremediation sites/factories. Although, plasmids with broad range metal resistance can be introduced as an alternative to genetically modified organisms as the application and use of latter in conventional sewage plants seems to be unrealistic.

5 Conclusion

The review demonstrates the potential applications of sustainable, ecofriendly techniques and strategies of treatment systems including biological (enzymatic and microbial) and combinations treatment to degrade textile dyes and dyeing effluents.

The type of technologies employed depends upon the composition of the effluents, types of dyed to be metabolized, and their side effects. Enzymatic processes are very promising green chemistry concepts for decolorisation/degradation of dyeing effluents. The enzymatic approach resulted in savings of dyeing additives and auxiliary chemicals. However, recently it has been seen that the cost for enzyme decolorisation/transformation is decreasing due to production of new ecofriendly technologies including genetic methods. The application processes involving microbial/bacterial cell cultures on an industrial scale allow complete mineralization/detoxification of textile dyes and is being encouraged by BIO-WISE (Major UK Government Programme). The BIOCOL process using whole *S. putrefaciens* cells immobilized on an activated carbon, to treat colored textile wastes have already been started by Northern Ireland.

However, it is very important to understand the mechanism of degradation/bioconversion/biotransformation of textile wastes, so that correct and specific treatment technologies can be employed. The biological and chemical techniques coupled with latest advances in genomics provide alternative and best treatments systems for dye decolorisation and are revolutionizing various aspects of biological sciences with direct implementation at industrial level. Indeed the basic understanding of gene profiling of specific single bacterial cultures will facilitate better degradation of azo dyes under anaerobic conditions. Moreover, the funding in this area of research should increase which will perhaps increases the use of bioprocessing solutions to the problem of colored wastewater in the textile industrial sectors.

Acknowledgements Financial support provided by University Grants Commission, Govt. of India; New Delhi through Maulana Azad National Fellowship (MANF) for **Sabiyah Akhter** is highly acknowledged.

References

1. Rather LJ, Islam S, Akhter S, Hassan QP, Mohammad F (2017) Chemistry of plant dyes: applications and environmental implications of dyeing processes. Curr Environ Eng 4:103–120
2. Rather LJ, Islam S, Shabbir M, Bukhari MN, Shahid M, Khan MA, Mohammad F (2016) Ecological dyeing of Woolen yarn with *Adhatoda vasica* natural dye in the presence of biomordants as an alternative copartner to metal mordants. J Environ Chem Eng 4:3041–3049
3. Ahlstrom L, Eskilsson CS, Bjorklund E (2005) Determination of banned azo dyes in consumer goods. Trend Anal Chem 24:49–56
4. Pierce J (1994) Colour in textile effluents—the origins of the problem. J Soc Dyers Col 110:131–134
5. Shahid M, Islam S, Mohammad F (2013) Recent advancements in natural dye applications: a review. J Clean Prod 53:310–331
6. Ibrahim NA, Moneim NMA, Halim ESA, Hosni MM (2008) Pollution prevention of cotton-cone reactive dyeing. J Cleaner Prod 16:1321–1326
7. Hu TL, Wu SC (2001) Assessment of the effect of azo dye RP2B on the growth of a nitrogen fixing cyanobacterium *Anabaena* sp. Biores Technol 77:93–95
8. Razo-Flores E, Donlon B, Lettinga G, Field JA (1997) Biotransformation and biodegradation of N substituted aromatics in methanogenic granular sludge. FEMS Microbiol Rev 20:525–538

9. Benigni R, Giuliani A, Franke R, Gruska A (2000) Quantitative structure-activity relationships of mutagenic and carcinogenic aromatic amines. Chem Rev 100:3697–3714

10. Rafii F, Franklin W, Cerniglia CE (1990) Azoreductase activity of anaerobic bacteria isolated from human intestinal microflora. Appl Environ Microbiol 56:2146–2151

11. Chung KT, Stevens SE Jr (1993) Degradation of azo dyes by environmental microorganisms and helminths. Environ Toxicol Chem 12:2121–2132

12. Hillenbrand T (1999) Die Abwassersituation in der deutschen Papier-, Textil- und Lederindustrie. Gwf Wasser Abwasser 14:267–273

13. Abadulla E, Tzanov T, Costa S, Robra KH, Cavaco-Paulo A, Gübitz GM (2000) Decolorization and detoxification of textile dyes with a laccase from *Trametes hirsuta*. Appl Environ Microbiol 66:3357–3362

14. Zhang X, Stebbing DW, Saddler JN, Beatson RP, Kruus K (2000) Enzyme treatments of the dissolved and colloidal substances present in mill white water and the effects on the resulting paper properties. J Wood Chem Technol 20:321–335

15. Knackmus HJ (1996) Basic knowledge and perspectives of bioelimination of xenobiotic compounds. J Biotechnol 51:287–295

16. Keck AA, Klein J, Kudlich M, Stolz A, Knackmuss HJ, Mattes R (1997) Reduction of azo dyes by redox mediators originating in the naphtalenesulfonic acid degradation pathway of *Sphingomonas* sp. Strain BN6. Appl Environ Microbiol 63:3684–3690

17. Nigam P, McMullan G, Banat IM, Marchant R (1996) Decolourisation of effluent from the textile industry by a microbial consortium. Biotechnol Lett 18:117–120

18. Nigam P, Banat IM, Singh D, Marchant R (1996) Microbial process for the decolorization of textile effluent containing azo, diazo and reactive dyes. Process Biochem 31:435–442

19. Zheng Z, Levin RE, Pinkham JL, Shetty K (1999) Decolorization of polymeric dyes by a novel *Penicillium* isolate. Process Biochem 34:31–37

20. Nigam P, Marchant R (1995) Selection of a substratum for composing biofilm system of a textileeffluent decolourizing bacteria. Biotechnol Lett 17:993–996

21. Kudlich M, Bishop E, Knackmuss HJ, Stolz A (1996) Simultaneous anaerobic and aerobic degradation of the sulfonated azo dye Mordant Yellow 3 by immobilized cells from naphtalenesulfonate-degrading mixed culture. Appl Microbiol Biotechnol 46:597–603

22. Shen CF, Miguez CB, Borque D, Groleau D, Guiol SR (1996) Methanotroph and methanogen coupling in granular biofilm under oxygen-limited conditions. Biotechnol Lett 18:495–500

23. Tan CG, Lettinga G, Field JA (1999) Reduction of the azo dye Mordant Orange 1 by methanogenic granular sludge exposed to oxygen. Biores Technol 67:35–42

24. Walker GM, Weatherly LR (1999) Biological activated carbon treatment of industrial wastewater in stirred tank reactors. Chem Eng J 75:201–206

25. Tse SW, Yu J (1997) Flocculation of *Pseudomonas* with aluminum sulfate for enhanced biodegradation of synthetic dyes. Biotechnol Tech 11:479–482

26. Banat IM, Nigam P, Singh D, Marchant R (1996) Microbial decolorization of textile dye containing effluents a review. Biores Technol 58:217–227

27. Martins MAM, Cardoso MH, Queiroz MJ, Ramalho MT, Campos AMO (1999) Biodegradation of azo dyes by the yeast *Candida zeylanoides* in batch aerated cultures. Chemosphere 38:2455–2460

28. Hao OJ, Kim H, Chiang PC (2000) Decolorization of wastewater. Crit Rev Environ Sci Tec 30:449–505

29. Slokar YM, Marechal AML (1998) Methods of decoloration of textile wastewaters. Dyes Pigm 37:335–356

30. Robinson T, McMullan G, Marchant R, Nigam P (2001) Remediation of dyes in textile effluent: a critical review on current treatment technologies with a proposed alternative. Biores Technol 77:247–255

31. Donlagic J, Levec J (1998) Does the catalytic wet oxidation yield products more amenable to biodegradation. Appl Catal B Environ 17:L1–L5

32. Kunz A, Reginatto V, Durán N (2001) Combined treatment of textile effluent using the sequence *Phanerochaete chrysosporium*—ozone. Chemosphere 44:281–287

33. Ledakowicz S, Solecka M, Zylla R (2001) Biodegradation, decolourisation and detoxification of textile wastewater enhanced by advanced oxidation processes. J Biotechnol 89:175–184
34. Pulgarin C, Invernizzi M, Parra S, Sarria V, Polania R, Püringer P (1999) Strategy for the coupling of photochemical and biological flow reactors useful in mineralization of biorecalcitrant industrial pollutants. Catal Today 54:341–352
35. Van der Bruggen B, De Vreese I, Vandecasteele C (2001) Water reclamation in the textile industry: nanofiltration of dye baths for wool dyeing. Ind Eng Chem Res 40:3973–3978
36. Bumpus JA (1995) Microbial degradation of azo dyes. Prog Ind Microbiol 32:157–176
37. Chung KT, Stevens SE Jr (1993) Degradation of azo dyes by environmental microorganisms and helminths. Environ Toxicol Chem 12:2121–2132
38. Farhadian M, Vachelard C, Duchez D, Larroche C (2008) In situ bioremediation of monoaromatic pollutants in groundwater: a review. Biores Technol 99:5296–5308
39. Radhika V, Subramanian S, Natarajan KA (2006) Bioremediation of zinc using Desulfotomaculum nigrificans: bioprecipitation and characterization studies. Water Res 40:3628–3636
40. Field JA, Stams AJM, Kato M, Schraa G (1995) Enhanced biodegradation of aromatic pollutants in cocultures of anaerobic and aerobic bacterial consortia. Antonie Van Leeuwenhoek 67:47–77
41. Madigan MT, Martinko JM, Parker J (2003) Brock biology of microorganisms, 10th ed. Prentice-Hall Inc., Simon & Schuster/A Viacom Company, Upper Saddle River, New Jersey, USA
42. Chung KT, Stevens SEJ (1993) Degradation of azo dyes by environmental microorganisms and helminths. Environ Toxicol Chem 12:2121–2132
43. Claus H, Faber G, Koenig H (2002) Redox-mediated decolorization of synthetic dyes by fungal laccases. Appl Microbiol Biotechnol 59:672–678
44. Knackmuss HJ (1996) Basic knowledge and perspectives of bioelimination of xenobiotic compounds. J Biotechnol 51:287–295
45. Lourenco ND, Novais JM, Pinheiro HM (2001) Effect of some operational parameters on textile dye biodegradation in a sequential batch reactor. J Biotechnol 89:163–174
46. Panswad T, Iamsamer K, Anotai J (2001) Decolorization of azo-reactive dye by polyphosphate- and glycogen-accumulating organisms in an anaerobic-aerobic sequencing batch reactor. Biores Technol 76:151–159
47. Tan CG, Lettinga G, Field JA (1999) Reduction of the azo dye mordant orange 1 by methanogenic granular sludge exposed to oxygen. Biores Technol 67:35–42
48. Zimmermann T, Gasser F, Kulla HG, Leisinger T (1984) Comparison of two bacterial azoreductases acquired during adaptation to growth on azo dyes. Arch Microbiol 138:37–43
49. Blumel S, Stolz A (2003) Cloning and characterization of the gene coding for the aerobic azoreductase from Pigmentiphaga kullae k24. Appl Microbiol Biotechnol 62:186–190
50. Pasti-Grigsby MB, Pasczcynski A, Goscyczynski S, Crawford DL, Crawford RL (1992) Influence of aromatic substitution patterns on azo-dye degradability by Streptomyces spp and Phanerochaete chrysosporium. Appl Environ Microbiol 58:3605–3613
51. Cao W, Mahadevan B, Crawford DL, Crawford RL (1993) Characterization of an extracellular azo-dye oxidizing peroxidase from Flavobacterium sp. ATCC 39723. Enzyme Microb Technol 15:810–817
52. Duran N, Rosa MA, D'Annibale A, Gianfreda L (2002) Applications of laccases and tyrosinases (phenoloxidases) immobilized on different supports: a review. Enzyme Microbiol Technol 31:907–931
53. Glenn JK, Akileswaran L, Gold MH (1986) Manganese-Ii oxidation is the principal function of the extracellular manganese peroxidases from Phanerochaete chrysosporium. Arch Biochem Biophys 251:688–696
54. Heinfling A, Bergbauer M, Szewzyk U (1997) Biodegradation of azo and phthalocyanine dyes by Trametes versicolor and Bjerkandera adusta. Appl Microbiol Biotechnol 48:261–266
55. Parshetti GK, Kalme SD, Gomare SS, Govindwar SP (2007) Biodegradation of Reactive blue-25 by Aspergillus achraceus NCIM-1446. Biores Technol 98:3638–3642
56. Kapoor A, Viraraghavan T (1995) Biosorption—an alternative treatment option for heavy metal bearing wastewaters: a review. Biores Technol 53(3):195–206

57. Marcanti-Contato I, Corso CR, Oliveira JE (1997) Induction of physical paramorphogenesis in Aspergillus sp. Braz J Microbiol 28:65–67
58. Olguín EJ (2003) Phycoremediation: key issues for cost-effective nutrient removal process. Biotechnol Adv 22:1–91
59. Ruiz J, Alvarez P, Arbib Z, Garrido C, Barragan J, Perales JA (2011) Effect of nitrogen and phosphorus concentration on their removal kinetic in treated urban wastewater by Chlorella vulgaris. Int J Phytorem 13:884–896
60. Khalaf MA (2008) Biosorption of reactive dye from textile wastewater by nonviable biomass of Aspergillus niger and Spirogyra sp. Biores Technol 99:6631–6634
61. Aravindhan R, Rao JR, Nair BU (2007) Removal of basic yellow dye from aqueous solution by sorption on green alga *Caulerpa scalpelliformis*. J Hazard Mater 142:68–76
62. Marungrueng K, Pavasant P (2006) Removal of basic dye (Astrazon Blue FRGL) using macroalga Caulerpa lentillifera. J Environ Manage 78:268–274
63. Acuner E, Dilek FB (2004) Treatment of tectilon yellow 2G by Chlorella vulgaris. Proc Biochem 39:623–631
64. Karacakaya P, Kilic NK, Duyugu E, Donmez G (2009) Stimulation of reactive dye removal by cyanobacteria in media containing triacontrol hormone. J Hazard Mater 172:1635–1639
65. Chivukula M, Renganathan V (1995) Phenolic azo dye oxidation by laccase from Pyricularia oryzae. Appl Environ Microbiol 61:4374–4377
66. Duran N, Esposito E (2000) Potential applications of oxidative enzymes and phenoloxidase-like compounds in wastewater and soil treatment: a review. Appl Catal B Environ 28:83–99
67. Mester T, Tien M (2000) Oxidative mechanism of ligninolytic enzymes involved in the degradation of environmental pollutants. Int Biodeter Biodegr 46:51–59
68. Li K, Xu F, Eriksson KEL (1999) Comparison of fungal laccases and redox mediators in oxidation of a non-phenolic lignin model compound. Appl Environ Microbiol 65:2654–2660
69. Soares GMB, De Amorim MTP, Costa FM (2001) Use of laccase together with redox mediators to decolourize Remazol Brilliant Blue R. J Biotechnol 89(2–3):123–129
70. Berry DF, Francis AJ, Bollag JM (1987) Microbial metabolism of homocyclic and heterocyclic aromatic compounds under anaerobic conditions. Microbiol Rev 51:43–59
71. Commandeur ICM, Parsons JR (1990) Degradation of halogenated aromatic compounds. Biodeg 1:207–220
72. Smith MR (1990) The biodegradation of aromatic hydrocarbons by bacteria. Biodeg 1:191–206
73. Kudlich M, Keck A, Klein J, Stolz A (1997) Localization of the enzyme system involved in anaerobic reduction of azo dyes by *Sphingomonas* sp. Strain BN6 and effect of artificial redox mediators on the rate of azo-dye reduction. Appl Environ Microbiol 63:3691–3694
74. Beydilli MI, Pavlostathis SG, Tincher WC (1998) Decolorization and toxicity screening of selected reactive azo dyes under methanogenic conditions. Water Sci Technol 38:225–232
75. Bromley-Challenor KCA, Knapp JS, Zhang Z, Gray NCC, Hetheridge MJ, Evans MR (2000) Decolorization of an azo dye by unacclimated activated sludge under anaerobic conditions. Water Res 34:4410–4418
76. Gingell R, Walker R (1971) Mechanism of azo reduction by Streptococcus faecalis II. The role of soluble flavins. Xenobiotica 1:231–239
77. Stolz A (2001) Basic and applied aspects in the microbial degradation of azo dyes. Appl Microbiol Biotechnol 56:69–80
78. Russ R, Rau J, Stolz A (2000) The function of cytoplasmic flavin reductases in the reduction of azo dyes by bacteria. Appl Environ Microbiol 66:1429–1434
79. Rau J, Stolz A (2003) Oxygen-insensitive nitroreductases NfsA and NfsB of Escherichia coli function under anaerobic conditions as lawsone-dependent azo reductases. Appl Environ Microbiol 69:3448–3455
80. Dos Santos AB, Cervantes FJ, Yaya-Beas RE, Van Lier JB (2003) Effect of redox mediator, AQDS, on the decolourisation of a reactive azo dye containing triazine group in a thermophilic anaerobic EGSB reactor. Enzyme Microbiol Technol 33:942–951
81. Yoo ES (2002) Chemical decolorization of the azo dye CI Reactive Orange 96 by various organic/inorganic compounds. J Chem Technol Biotechnol 77:481–485

82. Van der Zee FP, Bisschops IAE, Blanchard VG, Bouwman RHM, Lettinga G, Field JA (2003) The contribution of biotic and abiotic processes during azo dye reduction in anaerobic sludge. Water Res 37:3098–3109

83. Albuquerque MGE, Lopes AT, Serralheiro ML, Novais JM, Pinheiro HM (2005) Biological sulphate reduction and redox mediator effects on azo dye decolourisation in anaerobic–aerobic sequencing batch reactors. Enzyme Microbiol Technol 36:790–799

84. Dos Santos AB, Bisschops IAE, Cervantes FJ, Van Lier JB (2005) The transformation and toxicity of anthraquinone dyes during thermophilic (55 °C) and mesophilic (30 °C) anaerobic treatments. J Biotechnol 15:345–353

85. Lee YH, Matthews RD, Pavlostathis SG (2005) Anaerobic biodecolorization of textile reactive anthraquinone and phthalocyanine dyebaths under hypersaline conditions. Water Sci Technol 52:377–383

86. Lee YH, Matthews RD, Pavlostathis SG (2006) Biological decolorisation of reactice anthraquinone and phthalocyanine dyes under various oxidation–reduction conditions. Water Environ Res 78:156–169

87. Fredrickson JK, Kostandarithes HM, Li SW, Plymale AE, Daly MJ (2000) Reduction of Fe(III), Cr(VI), U(VI), and Tc(VII) by Deinococcus radiodurans R1. Appl Environ Microbiol 66:2006–2011

88. Lovley DR, Fraga JL, BluntHarris EL, Hayes LA, Phillips EJP, Coates JD (1998) Humic substances as a mediator for microbially catalyzed metal reduction. Acta Hydroch Hydrob 26:152–157

89. Keck A, Rau J, Reemtsma T, Mattes R, Stolz A, Klein J (2002) Identification of quinoide redox mediators that are formed during the degradation of naphthalene-2-sulfonate by Sphingomonas xenophaga BN6. Appl Environ Microbiol 68:4341–4349

90. Dos Santos AB, Bisschops IAE, Cervantes FJ, Van Lier JB (2004) Effect of different redox mediators during thermophilic azo dye reduction by anaerobic granular sludge and comparative study between mesophilic (30 °C) and thermophilic (55°C) treatments for decolourisation of textile wastewaters. Chemosphere 55:1149–1157

91. Walker R, Ryan AJ (1971) Some molecular parameters influencing rate of reductions of azo compounds by intestinal microflora. Xenobiotica 1:483–486

92. Moir D, Masson S, Chu I (2001) Structure-activity relationship study on the bioreduction of azo dyes by Clostridium paraputrificum. Environ Toxicol Chem 20:479–484

93. Gottschalk G, Knackmuss HJ (1993) Bacteria and the biodegradation of chemicals: achieved naturally, by combination, or by construction. Angew Chem Int Ed 32:1398–1408

94. Zimmermann T, Gasser F, Kulla HG, Leisinger T (1984) Comparison of two bacterial azoreductases acquired during adaptation to growth on azo dyes. Arch Microbiol 138:37–43

95. Knackmus HJ (1996) Basic knowledge and perspectives of bioelimination of xenobiotic compounds. J Biotechnol 51:287–295

96. Chang JS, Kuo TS (2000) Kinetics of bacterial decolorization of azo dye with Escherichia coli NO3. Biores Technol 75:107–111

Printed in the United States
By Bookmasters